ST. MARY'S COLLEGE OF MARYLAND
ST. MARY'S CITY, MARYLAND

PATTERNS OF HUMAN MOTION
a cinematographic analysis

PATTERNS OF

ARTISTS
Don Curtis, Robiny Case, Judy Pihl, John McLaughlin

PHOTOGRAPHER (tracing reductions)
Louis Musante

HUMAN MOTION
a cinematographic analysis

STANLEY PLAGENHOEF

University of Massachusetts

PRENTICE-HALL, INC., Englewood Cliffs, New Jersey

PATTERNS OF HUMAN MOTION
a cinematographic analysis
Stanley Plagenhoef

© 1971 by PRENTICE-HALL, INC., Englewood Cliffs, New Jersey

All rights reserved. No part of this book may be reproduced, in any form, by mimeograph or any other means, without permission in writing from the publisher.

13-654178-X

Library of Congress Catalog Card Number 75-135408

Printed in the United States of America

Current Printing (last digit)
10 9 8 7 6 5 4 3 2 1

PRENTICE-HALL INTERNATIONAL, INC., *London*
PRENTICE-HALL OF AUSTRALIA, PTY. LTD., *Sydney*
PRENTICE-HALL OF CANADA, LTD., *Toronto*
PRENTICE-HALL OF INDIA PRIVATE LIMITED, *New Delhi*
PRENTICE-HALL OF JAPAN, INC., *Tokyo*

CONTENTS

	Acknowledgments	ix
	Introduction	1
chapter 1	Methods of Data Collecting	3
chapter 2	Data Collecting	7
chapter 3	Anatomical Data	18
chapter 4	The Free Body Diagram for a One-Segment Motion	28
chapter 5	Two- and Three-Segment Motions	48
chapter 6	External Forces	58
chapter 7	Sports Equipment—Impact—Ball Spin	77
chapter 8	Throwing and Kicking	89
chapter 9	Analysis of Selected Sports	117
chapter 10	Maximum Joint Moments	152

appendix A	**Problems**		**159**
appendix B	**A Computer Program for the Kinetic Analysis of Human Motion**		**169**
appendix C	**Hand Calculation for Position 5**		**186**
appendix D	**Ball Bounce on Solid Surface Formulas**		**192**
appendix E	**Recommended Course Outlines**		**195**
	Bibliography		**199**
	Index		**223**

Tables

2-1	Individual Frame Exposure Changes (Seconds) Using a Variable Shutter Camera		8
3-1	Body Segments as Percentages of Total Body Weight (Dempster)		20
3-2	Specific Gravity (Dempster)		20
3-3	Segment Lengths as Percentages of Total Height (Dempster)		20
3-4	The Location of the Segment Center of Gravity from the Proximal and Distal Joint (r_p, r_d) and the Radius of Gyration from Each End (k_p, k_d). (From Dempster, B-7)		21
3-5	Body Segment Percentages of Total Body Weight for Living Men ($N = 35$)		25
3-6	Segment Length Percentages of Total Height for Living Men ($N = 35$)		25
3-7	Body Segment Percentages of Total Body Weight for Living Women ($N = 76$)		27
3-8	Segment Length Percentages of Total Height for Living Women		27
4-1	Data for Obtaining Body Segment Displacement		29
5-1	Data for Static Deep Knee Bend Problems		57
6-1	Ball and Striker Velocities Before and After Impact; Times of Impact		59
7-1	Swing and Impact Data on Sports Equipment		80
7-2	Properties of Balls		83
8-1	Kicking—Ball Velocities		101
8-2	Kicking—Foot Velocities Before Impact		101
8-3	Kicking—Striking Mass		102
8-4	Kicks Analyzed		102
10-1	Maximum Joint Moments for Exercises and Sports Motion Patterns		153
B-1	Number for Camera Angle Corrections		172
B-2	Computer Output		173
B-3	Computer Program		175

ACKNOWLEDGEMENTS

My teacher and friend, Wilfrid Dempster, guided me toward the mechanical analysis of human motion, and I'm sure realized before his death that he molded my life's work.

Robert Rosenbaum, Thomas Green, and Corrado Poli gave freely of their time, aiding me with the study of mechanics.

The many hours spent in writing the computer program by Myron Curtis and Gerald Calkins have advanced the field of kinesiology into a new era.

The anatomical data for women would not have been possible without many hours of work by Kirsti Kjeldsen, Janet Craft, and Margaret Toohey.

Although many of the motion pictures were taken during competition, many other projects resulted from special performances especially for my benefit. My thanks go to: Hubert Vogelsinger (soccer), Don Tonry (gymnastics), Robert Inferrer (karate), Steve Witkowski and his team members (bowling), Dwight Campbell and John Burke (golf), Phillip Moriarity and the many Yale and Wesleyan team members who gave their time (swimming), John Werle (basketball and volleyball), Marcia Smoke, Sperry Rademaker, Andy Toro, Roland Muhlen, Toby Cooper, Gerald Welborn, and dozens more (canoeing and kayaking), and Ed Serues, Paul Norton, and Carry Drakeford (squash).

James Laughnane coordinated the project on football injuries, and the tennis and golf projects were aided by Marcia Schneiweiss and Joe Gearon. The University of Massachusetts Audio-Visual Center and Photo Center not only made this monograph possible, but have enhanced my class teaching with many visual aids. Final comments and corrections which were very helpful were made by Judy and David Madsen and Richard Nelson.

PATTERNS OF HUMAN MOTION
a cinematographic analysis

INTRODUCTION

The recent changes that have taken place in the field of kinesiology were made possible by the collective work of many scholars and the technological advances of the past decade. The human body is a link system of variable lengths, weights, and shapes; by combining anatomical data, principles of mechanics, data-collecting equipment, and the computer, more specific and accurate data can be obtained on human movement than ever before.

Problems involving analysis of motion are complex because the body has many segments capable of moving in a multitude of directions simultaneously. It is important to recognize that all body motion is rotational, with each body segment revolving about one end or the other depending on whether the feet or the hands are stationary. Even though a segment may be moving in a straight line, the rectilinear motion is due to the rotation of two or more segments. The straight line movement of the trunk during a deep knee bend is thus the result of the combined rotations of the shank, thigh, and trunk.

Thirteen body segments are involved in the analysis of a complex three-dimensional, nonsymmetrical motion, such as a golf drive. Approximately 100 inertial forces, due to the motion of all the segments, must be calculated to obtain the moments of force at each joint. It is obviously necessary to learn to analyze human motion using simple motions before progressing to the complex motion of a golfer.

Even the analysis of a one- or two-segment motion requires a knowledge of mechanics so as to determine the forces due to rotation. An undergraduate course must fulfill this requirement. One semester each of high school algebra, physics, and trigonometry is sufficient to begin a college course in kinesiology. Anatomy should also be a prerequisite so that the lever system of bones and muscles is understood, and the joint centers can be readily located from external landmarks.

Introduction

The undergraduate kinesiology course should present the language of the kinesiologist, the pertinent anatomical data, the free body diagram, the basic laws of friction, ball restitution, and conservation of momentum. Students should also be introduced to the methods of obtaining data for analysis, along with the strengths and weaknesses of each method. They should learn the common language used by mathematicians, physicists, and engineers to describe position, direction, and movement, and should have available all the necessary anatomical data such as segment weights, centers of gravity, and radii of gyration. This course of study should culminate in the student's ability to draw a free body diagram showing all forces affecting a segment during a two-segment motion. To achieve this end it may be necessary to review trigonometric functions and problems dealing with torque, vectors, and static body positions.

A continued program at the graduate level should consider the more complex problems of human motion. Graduate students should be capable of drawing the free body diagram and calculating moments of force for a complete body motion. The bending, twisting, and elongating of the trunk present changes in the center of gravity and radius of gyration. Collecting data on complex motions becomes difficult and determination of inertial forces is practical only if a computer is used. Therefore a step-by-step procedure should be undertaken, proceeding from data collection to use of the computer programs necessary to obtain joint moments of force. A hand analysis of any simple motion will help the student to understand the need for the computer program. In addition to movement analysis, special related problems may be covered, such as: body impact, equipment impact (bat, racket, etc.), external continuous forces (rowing, bicycling), coefficient of restitution changes with velocity changes, ball-spin problems, and aerodynamic and fluid-dynamic problems.

Past approaches have emphasized verbal descriptions of movement based on the direction of each segment-movement and on listing the muscles that initiate the motion. After analyzing a few simple motions the inadequacies of this approach can be readily seen, because muscle functions are completely unpredictable when only the direction of segment motion is known. It is only by obtaining data on displacement, angular velocity and acceleration, and moments of force that a motion can be interpreted. These complexities have been realized in the past, but a lack of application of the proper techniques has inhibited progress toward their complete understanding. People making prosthetic devices have applied the mechanics necessary for a two-segment movement, but the computer now allows the practical extension of mechanics to the analysis of a complex whole body motion. Whether the need is for a simple motion in rehabilitation or a complex movement in sports, the proper application of the available techniques is essential to the upgrading of the analysis of human motion. Hopefully, we are leaving the era of indiscriminate muscle listing and are moving on to collecting data for a more critical analysis.

The following chapters present methods of obtaining and analyzing data on human motion. In addition, a bibliography is provided. It is divided into sections (A through J) to facilitate finding references of similar content as well as source materials. References to these lettered and numbered sources are made throughout the text.

chapter 1

METHODS OF DATA COLLECTING

The kinetic analysis of human motion has been limited in the past due to the complex patterns of motion that a human is capable of performing. The development of data-collecting equipment, together with the isolated work of many individuals, have given the field of kinesiology the tools needed for kinetic analysis. The electromyograph, electrogoniometer, stroboscope, high- and normal-speed motion picture cameras, X-ray, force plate, force transducer, and the computer have all contributed to the more precise recording and analysis of motion. Methods of recording motion, together with anatomical data that reduce the body to a link system of known weights and lengths, now make it possible to obtain a great deal more information than ever before.

So many people have contributed to the development of the field that recognition cannot be given to them all in a nonhistorical text such as this. Any in-depth study of the related literature is left to the individual, with substantial aid being given by the selected Bibliography at the end of this text.

ELECTROMYOGRAPHY

The use of electromyography (EMG) presents a method for obtaining comparative data relating to the force of contraction of individual muscles. Surface or needle electrodes are placed as close as possible to the motor unit controlling an individual muscle, and the extent of muscular contraction is recorded on moving graph paper or on an oscilloscope. The closer the placement of the electrode to the motor unit, the more intense the reading transmitted to the recording device. Therefore, the pattern recorded becomes more important than its intensity for interpreting the extent of the force of contraction of the group of fibers involved.

There is no other method for recording the extent of muscle contraction at any moment during a movement. EMG can answer questions such as: Does the biceps femoris contract more during a deep knee bend or during bicycle riding? Does the pectoralis major contract more during a supine press using 100 lb, or during a supine straight-arm lateral raise using 15 lb? What is the extent of contraction of hip flexors during straight-leg and bent-leg sit-ups? The ability to answer these questions is largely dependent upon the nearness of the motor unit to the skin surface. Some conclusions obtained by different investigators show that:

(1) The brachialis flexes the forearm in all positions and recruits the biceps brachii as the load increases;
(2) The erector spinae remain relaxed in the initial stages of heavy lifting if the lower back is curved;
(3) The medial head of the triceps is the prime extensor of the elbow, and then the lateral and long heads are recruited as the load is increased.

The EMG has several limitations. Besides being costly, the number of channels available limits the number of simultaneous readings. Also, movement must be limited to the confines of the laboratory due to the restrictions of the attached wires, although the advances being made in telemetry make the equipment much more versatile. Difficulty in locating motor points, inconsistent pick-up and print-out, and external interference often produce questionable data. The technique requires experience and an ability to interpret data properly.

The equipment used and the interpretation and accuracy of the results obtained are fully presented by Basmajian (A-3). A partial list of numerous projects using EMG is available in sections A, H, and I of the Bibliography (A: 1–5, 8, 14, 22, 33, 36, 43, 45, 49, 50, 56, 58, 59, 62, 64, 66, 68, 69; H-4; I: 33, 39, 50, 54). The use of EMG, its data and limitations, and its possible use in conjunction with other data-collecting methods should be presented; however, further discussion in this text will be limited to predicting EMG readings of certain muscles when inertial joint forces are determined mathematically.

ELECTROGONIOMETRY

Karpovich and Karpovich (A-39) designed the electrogoniometer to obtain instant records of joint angle changes regardless of the position of the segments measured. The equipment's advantages of speed of recording and obtaining angular changes during a three-dimensional movement make this the only means of measuring the actual elbow angle during the arm stroke of the crawl, the knee angle during a soccer kick even though the whole body is revolving about the nonkicking foot, or the hip-shoulder rotation during a trunk twisting movement. It has the advantage of obtaining a fast, continuous record of simple motions that would take a longer time to record using other methods.

The limitations are very similar to those of electromyography. The laboratory set-up, with wires attached, limits the possible movements that can be analysed. The number of joints that can be measured simultaneously is limited by the equipment available. A kinetic analysis can be done only if the motion measured is planar and if all joints in motion are recorded; therefore, the data obtained cannot be used for a kinetic analysis during a three-dimensional motion. This places extreme limitations on use of the equipment. Ricci (A-54) illustrates the equipment and describes its use, and section A of the Bibliography lists several studies that have used the electrogoniometer.

STROBOSCOPE

Stroboscopic pictures were made famous by Edgerton and Killian (A-19) with their photographic studies of the multi-positions of a golf swing, the position sequence of a springboard diver, and the impact of golf and tennis balls. Stroboscopic pictures produce a composite picture of a whole body motion with a minimum amount of time and effort by recording several instantaneous body positions on one negative of a still camera with an open shutter. The number of positions recorded is controlled by the frequency setting of the flashing light. If a planar motion is recorded, it is possible to obtain body segment displacement. The measured data of strobe pictures and electrogoniometer readings should produce the same results if obtained simultaneously. If the movement is three-dimensional, the electrogoniometer continues to measure actual joint angles, whereas strobe pictures make it possible to obtain angular displacement in the plane of the picture as determined by the camera placement. This is useful in determining the angular velocities, angular accelerations, and forces that contribute to ball velocity in golf, tennis, baseball, etc., if the picture is taken in the plane of the ball flight.

Thus, use of the data of strobe pictures varies from that of the electrogoniometer when a three-dimensional motion is analyzed. The former method has the advantage of obtaining data without the restrictions of attached wires. Another advantage of strobe lighting is the ability to obtain impact pictures with it. A picture at impact can be obtained by setting off the flash by sound, or the object to be struck can be wired so that the picture is taken at contact.

There are several disadvantages to using strobe techniques. An open camera during the flashing period requires a dark room. Therefore, the movements to be analyzed are restricted by the lab set-up, and their performance is hampered by darkness. A revolving disc and other techniques have been employed to obtain strobe pictures in daylight, but the results have been disappointing. The biggest disadvantage is the overlapping image of a whole body motion when the performer remains in one spot. This means that the path of the joint centers cannot be distinguished and an analysis is thus impossible. Because strobe lighting is useful only if the body changes position sufficiently to distinguish joint centers, the technique is very limited for practical use in analyzing motion.

MOTION PICTURES

The motion picture camera makes it possible to record the performance of a subject without his knowledge, under actual competitive conditions, and permits kinetic analysis of any type of motion. High-speed cameras also measure the time of impact of colliding bodies, and make it possible to solve problems of friction and ball spin. Motion pictures can be used in conjunction with electromyography by comparing the moments of force obtained with the EMG readings of specific muscles.

Because motion pictures make it possible to analyze almost any body motion, this technique is utilized almost exclusively in this text, and its problems, techniques, and utilization will be discussed in detail. There are some disadvantages to using motion pictures that should be noted here. Taking movies indoors requires floodlights, and this interferes with performance to some extent. The main objection would probably be the time it takes to make a composite tracing from frame-by-frame projections. However, the time spent is generally worth the effort, as a complete kinetic analysis produces rewarding results.

MEASURING EXTERNAL FORCES

Equipment for measuring external forces has further advanced motion analysis. External forces due to impact or continually changing forces due to pushing and pulling (rowing) must be measured if a kinetic analysis of body motion is to be made. The force transducer can obtain the necessary data for such actions as: pulling on an oar, pushing on a bicycle pedal, or striking a tennis ball (actual force of impact transmitted to the hand through the racket). Ishiko (I-44) has used this technique in rowing.

The force plate has also been a useful addition to motion analysis. Brouha (A-9) illustrates the equipment and describes its use. A force plate measures the vertical and horizontal forces in one plane exerted on a floor plate. It can be used to check the accuracy of a kinetic analysis which uses motion pictures if the plane of the camera is placed to correspond with the force plate's horizontal direction measurement. Many studies pertaining to gait have used the force plate, and it is used in designing prosthetic devices to measure gait variations.

THE COMPUTER

The application of the high-speed computer to kinetic analysis has been the greatest recent advance in the field of kinesiology. Even though a motion is recorded, anatomical data applied, measurements made, and formulas available for application, the remaining work is monumental and thus prohibitive. The computer has cut 20 hours of slide rule work to one minute, and has made the analysis of a complex whole body motion practical. Computer programming has made it possible to obtain body segment angular velocities and accelerations, horizontal and vertical forces, joint moments of force, total body centers of gravity, and the contribution of each body segment to the whole motion. By noting the points of maximum velocity, acceleration, and deceleration of each body segment, the proper timing of one body segment relative to the next indicates the efficiency of the movement. The application of these data to specific sports, and to similar patterns of motion, bridges the gap between theory and practice.

chapter 2

DATA COLLECTING

In this text the recording of a whole body motion for analysis is done only by motion pictures. Motion pictures make it possible to record a motion during competition without subject awareness. They are also the only method available that allows a total body force and joint moment analysis of all types of motions, while also permitting a detailed and unobstructed view of any instantaneous position. A motion is recorded in a selected plane, determined by the camera position, so the forces contributing to the velocity of a ball thrown or hit may be isolated. Two or three cameras may be used to obtain simultaneous records of a motion from different angles, and thus permit calculation of forces and moments in three dimensions. The method chosen for data collection must answer best the pertinent questions about body motion. Because a vast number of motions can best be analyzed using motion pictures, the remainder of this text specializes in this technique.

OBTAINING MOTION PICTURES

The number of positions recorded by the motion picture camera of a whole body motion is determined by the frame-per-second setting. A slow motion, such as swimming, requires only 24 frames per second (f/s) and a fully open shutter setting. Most fast motions, such as a tennis serve or a baseball pitch, require at least 64 f/s and a $\frac{1}{4}$ open shutter. A faster motion, such as that of a ball after impact in a golf drive, requires about 80 f/s and a $\frac{1}{4}$ open shutter. It is important that the camera have a variable shutter setting to reduce the time exposure of each frame. The changes in exposure due to the frames per second setting and variable shutter settings are shown in Table 2-1 (specifically for the Pathe Professional Reflex 16 mm Model).

Table 2-1 Individual Frame Exposure Changes (Seconds)
Using a Variable Shutter Camera

FRAMES/SECOND	VARIABLE SHUTTER		
	Fully Open	$\frac{1}{2}$ Open	$\frac{1}{4}$ Open
16	$\frac{1}{32}$	$\frac{1}{64}$	$\frac{1}{128}$
24	$\frac{1}{48}$	$\frac{1}{96}$	$\frac{1}{192}$
32	$\frac{1}{64}$	$\frac{1}{128}$	$\frac{1}{256}$
48	$\frac{1}{96}$	$\frac{1}{192}$	$\frac{1}{384}$
64	$\frac{1}{128}$	$\frac{1}{256}$	$\frac{1}{512}$
80	$\frac{1}{160}$	$\frac{1}{320}$	$\frac{1}{640}$

The following problems will help determine the camera settings for various sports:

Problem 1 A motion picture camera set at 80 f/s at $\frac{1}{4}$ open shutter has a frame exposure of $\frac{1}{640}$ sec. What ball velocity would be limited to a 4 in. blur? X = velocity of ball. A ball going 640 ft/sec will go 1 ft in $\frac{1}{640}$ sec; therefore,

$$\frac{X \text{ ft}}{1 \text{ sec}} = \frac{4 \text{ in.}}{\frac{1}{640}} \qquad X = 213.3 \text{ ft/sec}$$

Problem 2 If the camera setting produces a frame exposure of $\frac{1}{384}$ sec as determined from Table 2-1, how long is the blur if the ball speed is 192 ft/sec? X = blur in inches.

$$\frac{192 \text{ ft}}{1 \text{ sec}} = \frac{X}{\frac{1}{384}} \qquad X = 6 \text{ in.}$$

It is important to determine the length of the blur of fast moving objects to make sure that the camera settings will give usable pictures. If a golf ball has a blur much longer than 4 in., the image cannot be seen, but if less than 4 in. the blur is measurable. When measuring the velocity of a golf club head or a badminton racket head, the blur must be reduced to less than 4 in. If the blur were to be reduced more, it would require specialized high-speed equipment. Such a camera is more expensive, but highly desirable for very fast motions and for problems involving impact. High-speed cameras also have built-in timing lights and various-size aperture slits.

Taking pictures outdoors does not present lighting problems with the fast films available today. Three steps must be followed so that the pictures can be analyzed.

(1) Place a meter stick and a hanging plumb bob in the plane of motion to obtain the scale and to provide a vertical reference when the film is projected;
(2) Take a picture of a sweep second hand clock to obtain the frames/second more accurately than the camera setting;
(3) Correct for camera error when the motion is three-dimensional.

Camera error is determined as shown in Fig. 2-1. The scaled drawing includes the camera-to-subject distance, width of field, and subject size. The percentage

Data Collecting

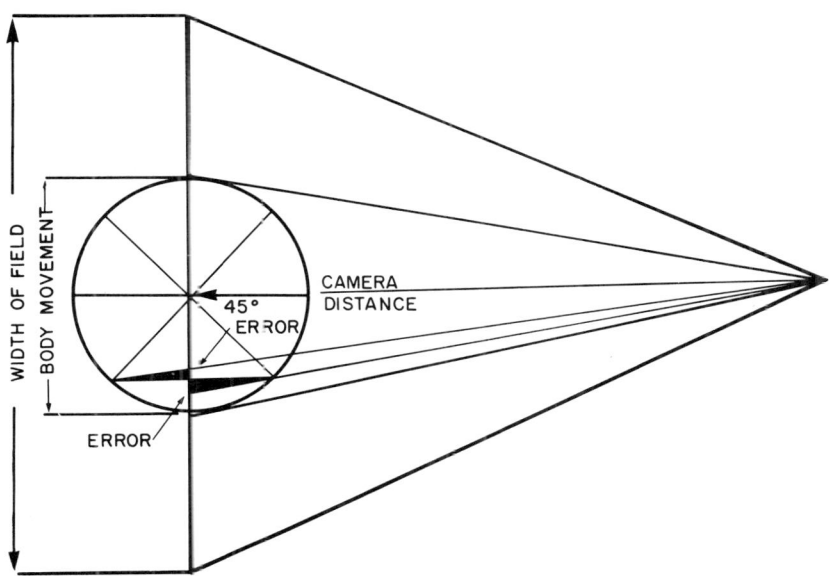

Fig. 2-1
The determination of the error in the body segment length as recorded by a camera as the camera distance lens, and image size and angle change.

change in the projected length of a line as seen by a camera is determined and subtracted from the measured lengths if the motion is in front of the plane, and added if the motion is behind the plane. There is less error as the focal length of the lens increases, and, therefore, a 50 mm lens should be used when possible.

Taking pictures indoors is more difficult due to the lighting problems. Good picture contrast is obtained when the subject is lighted with floodlights from about 30° forward of the plane of the picture on both the right and left sides, and a single floodlight is placed in line with the camera. Clear, well-lighted pictures of the proper size are necessary to locate joint centers. When possible, the subject should wear a minimal amount of clothing, and the planes of the joint centers should be marked with a skin pencil. If the background can be controlled, it should contrast sharply with the subject (e.g., skin against black velvet). Accuracy in the location of the joint centers will then depend on the experience of the analyst in locating the joint centers from external body landmarks.

CAMERA SETTINGS

The speed of the motion photographed controls camera settings by predetermining the frames per second and shutter settings. (Determine the time exposure for each frame based on the maximum velocities occurring during the motion. See Table 6-1 as an aid for estimating velocities.) The available light then dictates the lens setting and thus the depth of field. Using 4X film (ASA 400) and adjusting the processing can help decrease time exposure.

Data Collecting

The photographic information presented here is meant only to make you aware of the problems involved and the many variables that exist. The *American Cinematographer Manual* (A-47) and Hyser (A-32) cover most aspects of cinematography. They answer questions dealing with measurement error due to relative shutter-subject movement, velocity and acceleration of subject motion, lens aberrations, atmospheric refraction, film distortion, film-plane relation to object-plane, and the timing of the frames per second. Many types of equipment (cameras, lens, photographic materials), lighting, exposure, and additional related references are also presented. Because of the extensive and detailed coverage of pertinent motion picture problems included in these two references, the student should consult them for information about specific problems; this text will not discuss these areas further.

COMPOSITE TRACINGS

The composite tracing of motion pictures is the worksheet used for recording all data. The joint centers are marked and connected to produce a stick figure showing the positions of the body segments (Fig. 2-2). This is done using

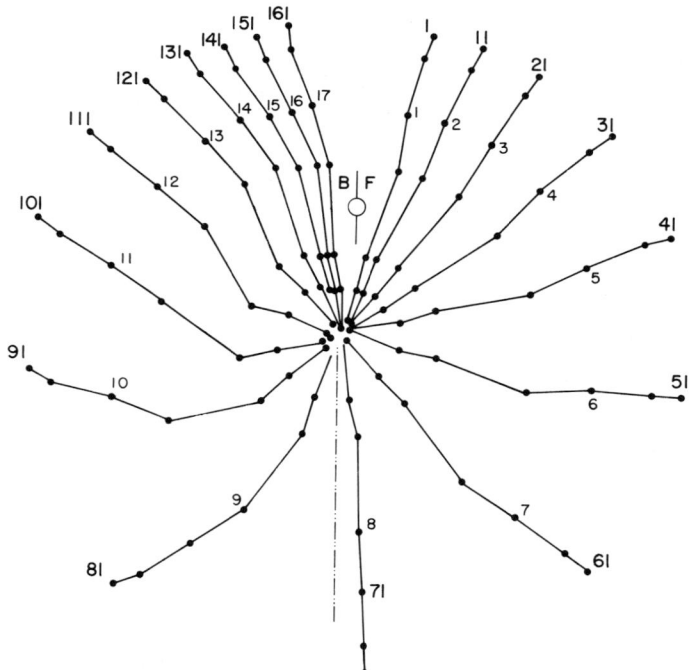

Fig. 2-2
A composite tracing with lines representing the body segments (Tonry front giant swing). Dots are placed at the wrist, elbow, glenohumeral, hip, knee, and ankle joints, and at the center of gravity of the feet. The motion is recorded at 64 f/s; every 10th frame is plotted.

11

Data Collecting

a Super Sports Analyst projector (L-W Photo Inc., Model 800) that has a special heat-resisting element so the film will not burn when stopped. Determining the joint centers introduces a subjective error which varies with the tracer's knowledge of anatomy. Aids for locating the joint centers from external landmarks are illustrated in Figs. 2-3 to 2-16.

Fig. 2-3

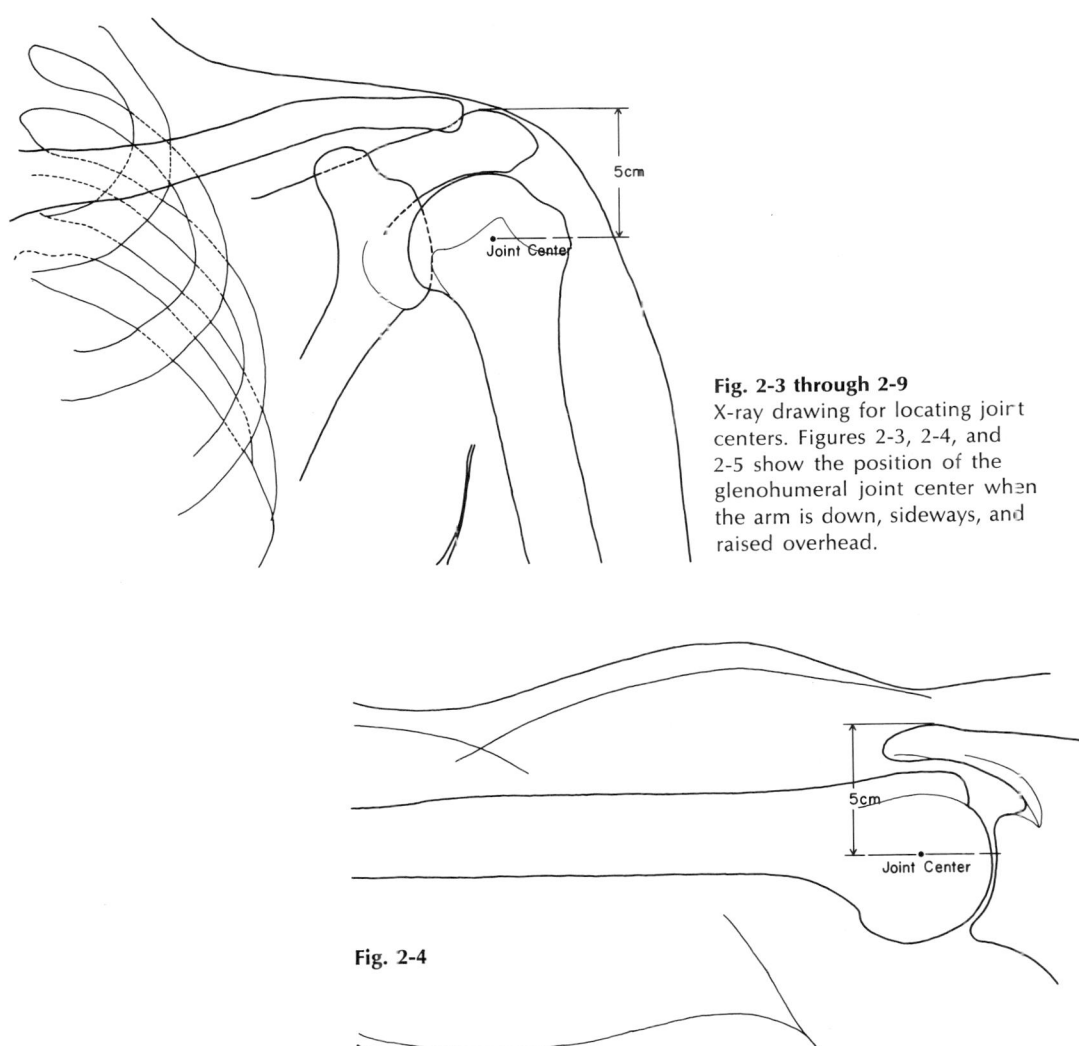

Fig. 2-3 through 2-9
X-ray drawing for locating joint centers. Figures 2-3, 2-4, and 2-5 show the position of the glenohumeral joint center when the arm is down, sideways, and raised overhead.

Fig. 2-4

Data Collecting

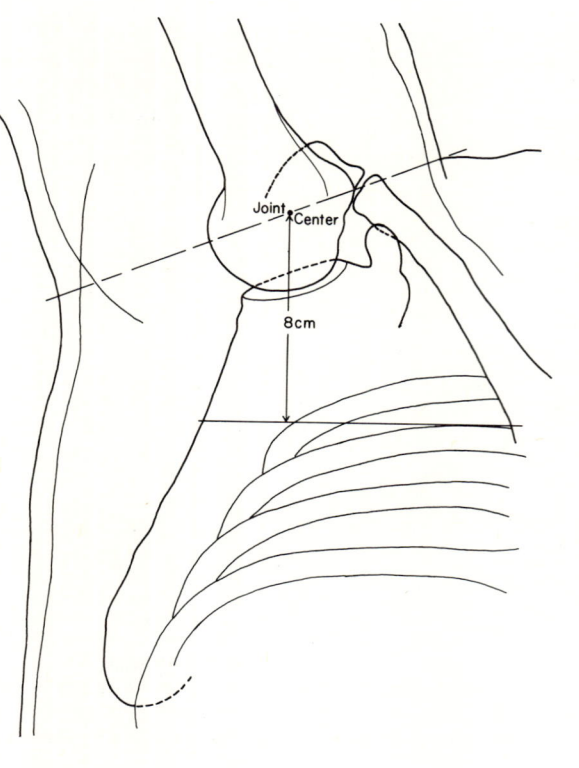

Fig. 2-5

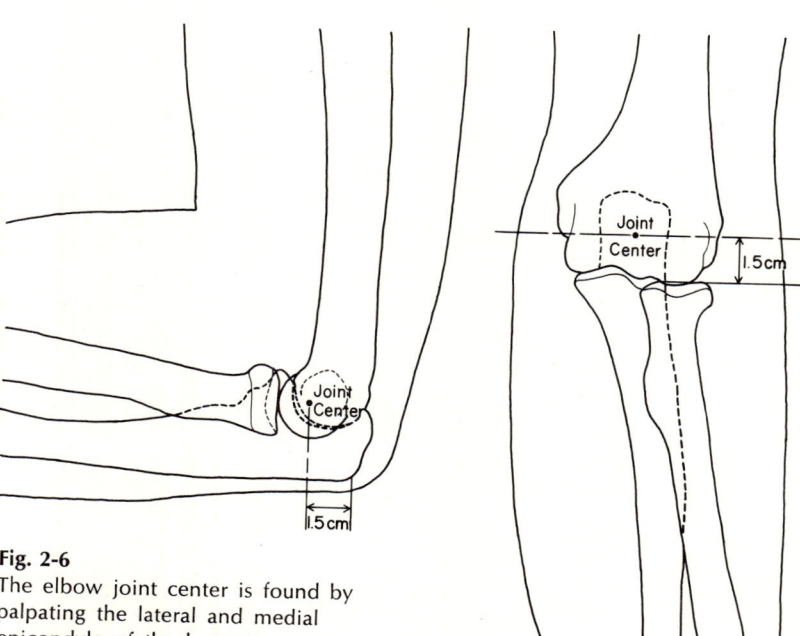

Fig. 2-6
The elbow joint center is found by palpating the lateral and medial epicondyle of the humerus.

13

Data Collecting

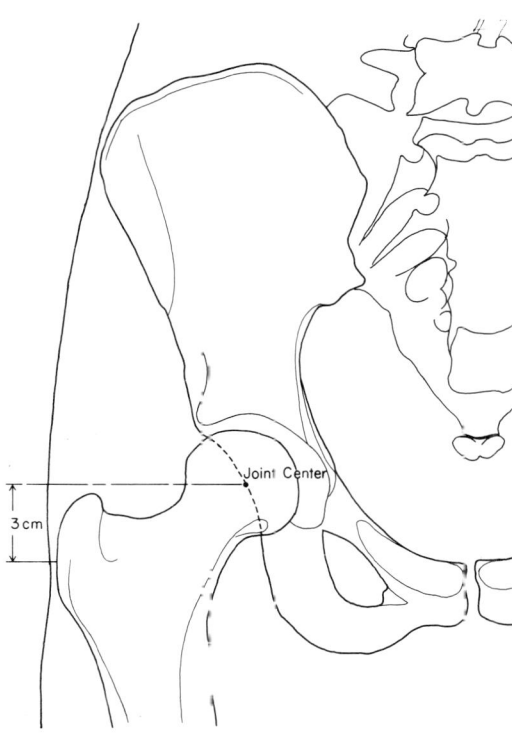

Fig. 2-7
The hip joint center is located approximately 3 cm above the most lateral bony prominence of the greater trocanter.

Figs. 2-8, 2-9
The knee center is located by determining the center of the flat portion of the condyles of the femur.

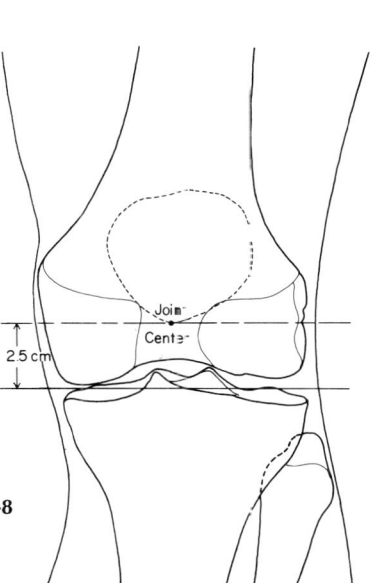

Fig. 2-8

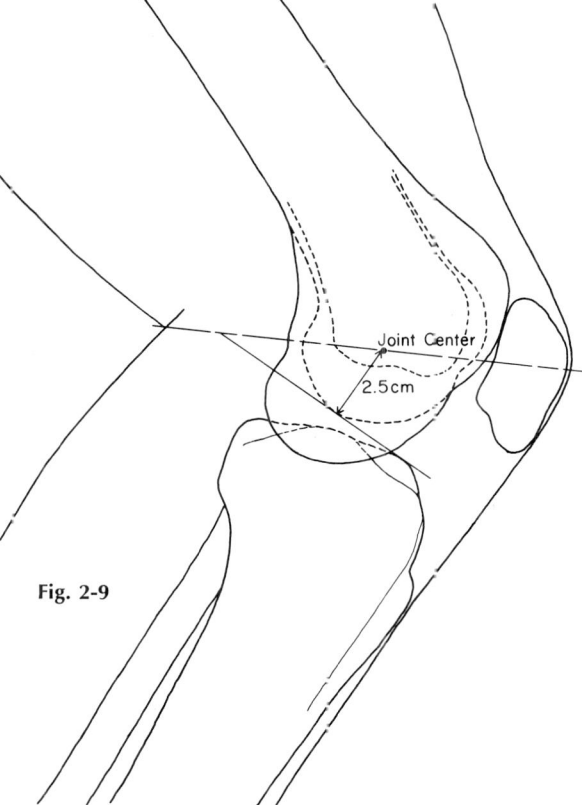

Fig. 2-9

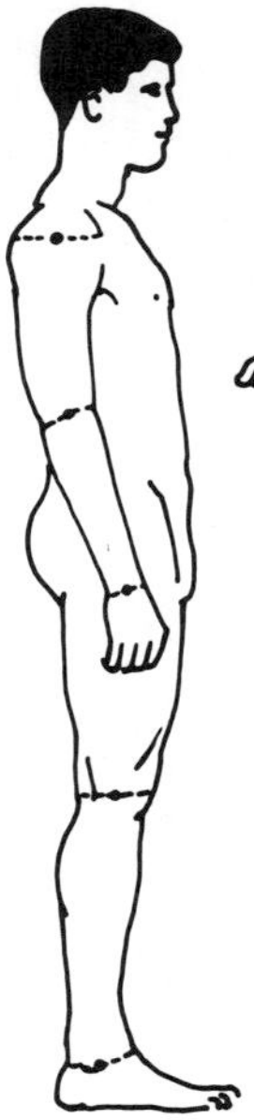

14

Data Collecting

Fig. 2-10

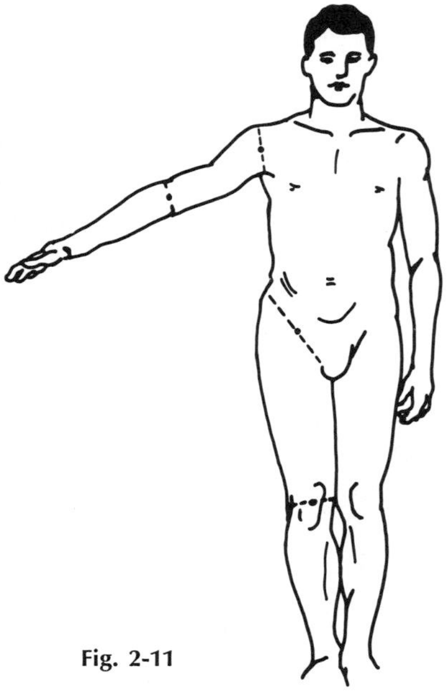

Fig. 2-11

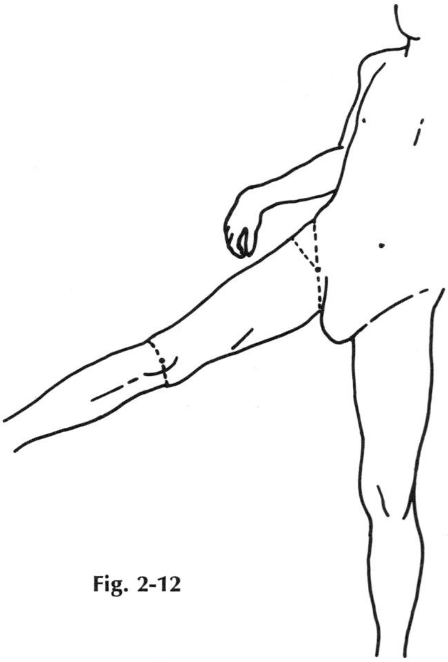

Fig. 2-12

Fig. 2-10 through 2-16
Skin creases and bony landmarks help locate the *plane* of the joint center so that the *position* of the center can be located. If a skin pencil is used to draw the plane of the joint center, the center must be located properly when the segment movement is out of the vertical plane (see Figures 2-14 through 2-16).

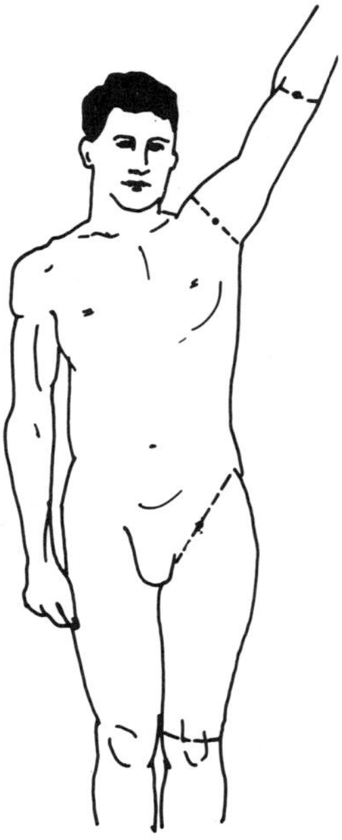

Fig. 2-13

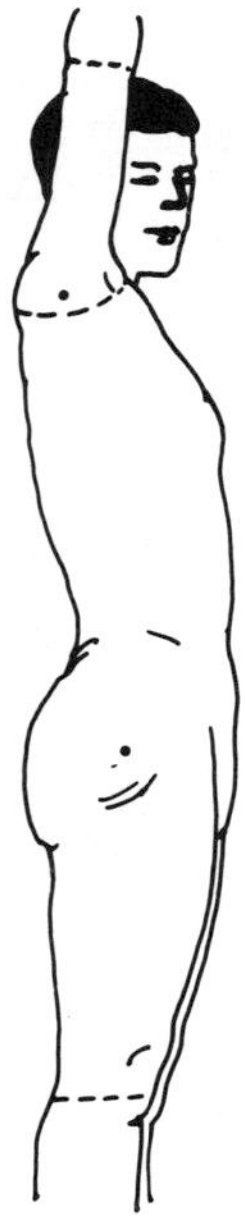

Fig. 2-14

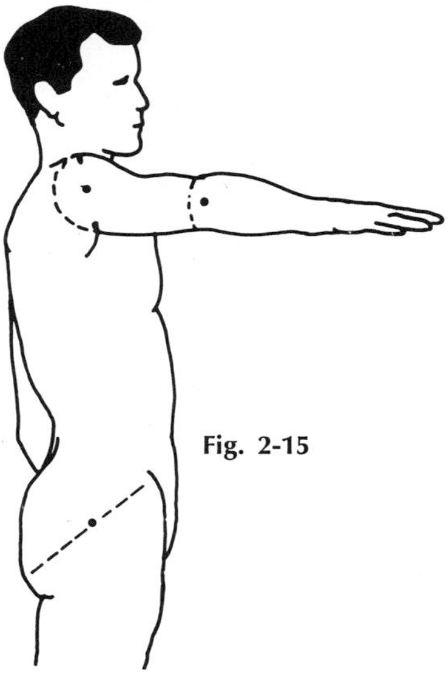

Fig. 2-15

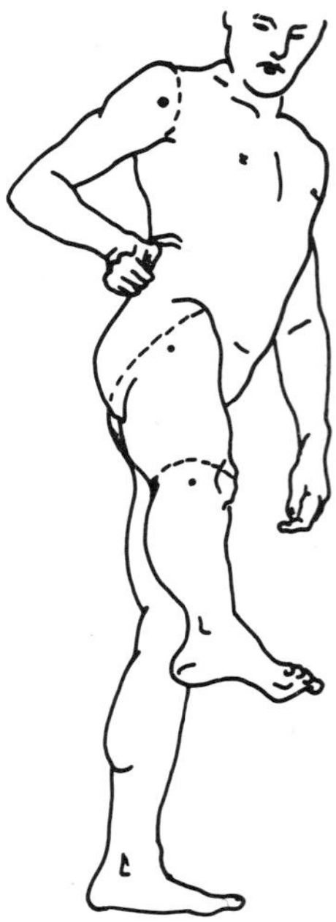

Fig. 2-16

15

Data Collecting

If the camera is perfectly rigid on a tripod, the tracing can be made without adjustments. Slight adjustments are usually necessary to keep the fixed part of the body (feet) in the same position as the film is advanced. It is wise to draw a fixed line from the background, so that the sheet may be adjusted should movement occur due to camera vibrations or due to slight projector deviations on advancement of the film. There are also situations where the camera is not held stationary during picture taking, such as a scuba diver taking pictures of a swimmer from the pool bottom, or panning with a rower so as to analyze the rower relative to the shell. A tracing table which makes it easy to adjust the tracing paper for each frame is shown in Fig. 2-17. The opaque paper on the glass top table can be moved freely and placed in the correct "base" position for each frame.

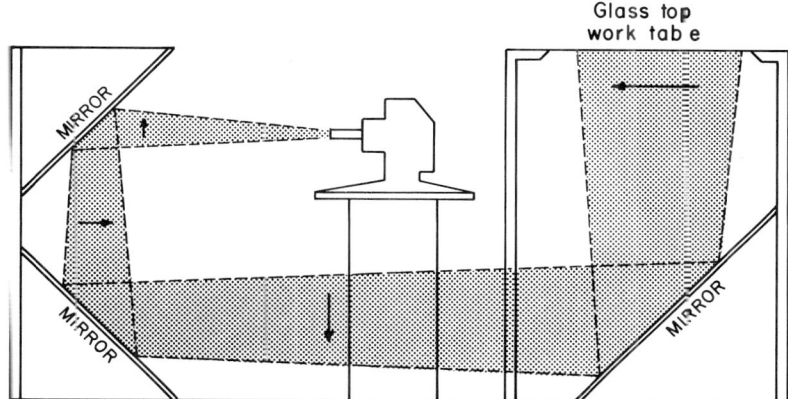

Fig. 2-17
Glass top tracing table. This equipment is used when the paper must be moved for each frame because of camera movement or in recording a body motion relative to a moving object (i.e., rowing, bicycling).

A Vanguard or Recordac film viewer can also be used, but the projected picture is small. Repeated measurements of one subject produces the best results when the scale of the projection is $\frac{1}{4}$ to $\frac{1}{5}$ life size. The ideal equipment should be capable of recording segment angles from the optimum size and taping the data for immediate use by the computer. The composite stick figure should also be computer printed, so that the total image of the movement is retained.

A computer-printed graph producing a line or point-by-point graph may be obtained using a 32K CDC 3600 computer with a card reader and printer. This program is used at the University of Massachusetts. (Programmer — Robert M. Janoski, University of Wisconsin). This program was utilized to produce the top view of a motion given the front and side views. This makes it possible to obtain top view pictures of many motions that would be impossible to obtain during actual competition (e.g., baseball batter) by using two ground level cameras placed appropriately at 90° to each other.

No matter how sophisticated the equipment becomes in the future, the student should do all of his tracings by hand. He will then realize the errors introduced into an analysis by reproducing a poor composite tracing, and gain experience in locating joint centers more accurately. After sufficient experience in tracing all types of motions has been logged, the computer may be used exclusively.

Data Collecting

A beginner should not get involved in the tracing and analysis of a complicated motion, so a classification of motion in a progressive order of difficulty of analysis is presented below. Motion can be classified according to the number of segments in motion, whether symmetrical or nonsymmetrical, the degrees of freedom (planar or three-dimensional), whether the trunk can be considered a rigid body, whether there is a fixed point throughout the entire motion, and whether there are external forces to be applied. The suggested order of difficulty is as follows:

I. No movement; analyze any static position:
   A. Holding the arm out straight forward
   B. Bending at the hips while standing
   C. Sitting up at 45°
II. Rotation of a single rigid body:
   A. Simple pendulum
   B. Forearm flexion (no wrist motion)
III. Symmetrical, planar motion:
   A. Two arm curl, analyze two segments only (upper and forearm)
   B. Two arm curl, continue the analysis to include all the nonmoving joints
   C. Total body motion (standing broad jump, parallel bar kip, giant swing, underhand basketball throw)
IV. One segment, nonplanar motion:
   A. Moving the straight arm upward and sideward at 45° to the plane of the camera
V. Symmetrical, nonplanar motion:
   A. Deep knee bend (knees move out of the plane)
   B. Swim start, butterfly and breast strokes (arms move out of the plane)
   C. Jumping jack (arms and legs moving in different planes when knees bend for the jump)
VI. Nonsymmetrical, planar motion:
   A. Walking (limit analysis to one side of the body)
VII. Nonsymmetrical, nonplanar motion:
   A. A fixed point throughout the motion (tennis serve — left foot fixed)
   B. A fixed point maintained after a preliminary ready position is taken (baseball batter, baseball pitcher — after front foot fixed)
   C. A fixed point attained in the middle of a motion (badminton smash, and lacrosse shot — jump from right foot to left foot during movement)
   D. Fixed point, link system moves up one leg and down the other (soccer kick)
   E. Fixed point, both sides of the body must be analyzed separately (golf swing)
   F. Fixed point, equipment moving (pole vault)
VIII. Trunk flexion, extension or shoulder elevation (the trunk cannot be considered as a rigid body):
   A. Sit-up
   B. Leg lift hanging from a bar
   C. Soccer heading (the trunk must be divided further into moving parts)
IX. External forces:
   A. Wall pulleys — forces on hands, feet fixed
   B. Sudden force application of impact (soccer heading, handball)

Data Collecting

    C. Continuously changing force:
       1. Rowing — force on hands, weight on seat, feet fixed
       2. Bicycling — force on feet, weight on seat, hands fixed
       3. Soccer heading — the movement of the arms can be considered as an external force to the link system taken from feet to head

X. No fixed point:
    A. Motion in a fluid (swimming)
    B. Motion in the air (high jump, basketball jump shot, soccer head)

chapter 3

ANATOMICAL DATA

Braune and Fischer (B-4) presented the first significant work on segment centers of gravity in 1889 based on their dissection of four cadavers. Their report presented data obtained by an anatomist, Harlass, but rejected it in favor of their own more extensive research. It was not until 1955 that more comprehensive work, based on the dissection of eight cadavers, was presented by Dempster (B-7). A monograph summarizing the work of Braune and Fischer, Amar, Fischer, and Dempster was presented by Krogman and Johnson (B-14) in 1963. A summary of Dempster's work and Braune and Fischer's work is also presented by Williams and Lissner (H-29). Dempster's chapter in *Biomechanical Studies of the Musculo-Skeletal System* (D-7) presents the locations of segment centers of gravity, segment densities, segment weights as a percentage of total body weight, the water displacement method for obtaining segment weights, and use of the free body diagram. This text uses only Dempster's data, and expands his work to make it possible to analyze complex body motions. The work of Clauser et al (B-5) is also an expansion of the techniques used by Dempster and presents interesting new data for future use.

Figure 3-1 indicates the planes of the joint centers, which were used by Dempster for the dissection of the eight cadavers. The same planes are marked on living subjects to obtain body segment weights using the water displacement method. Tables 3-1 through 3-4 present the anatomical data contributed by Dempster (B-7).

19
Anatomical Data

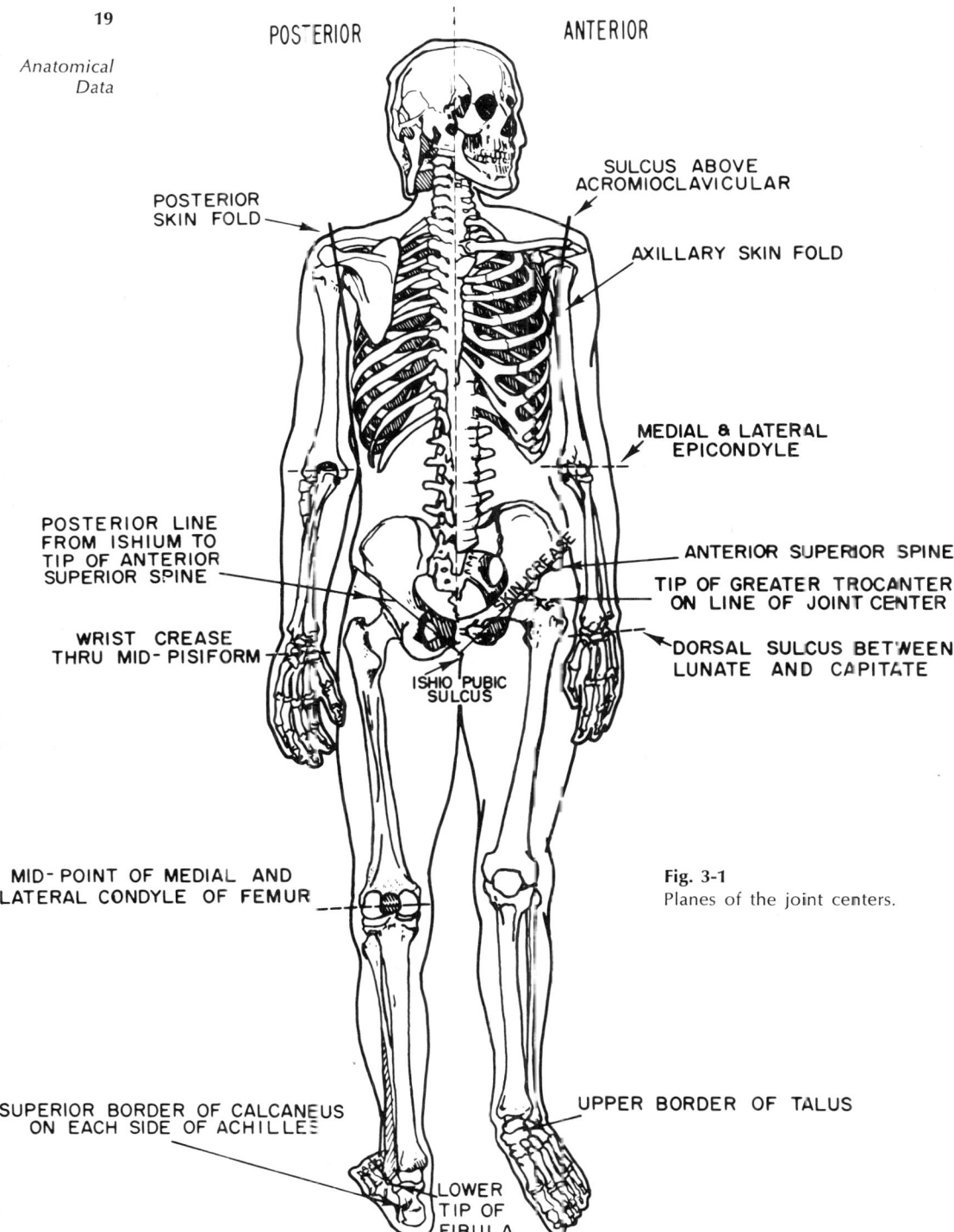

Fig. 3-1
Planes of the joint centers.

Table 3-1 Body Segments as Percentages of Total Body Weight (Dempster)

Hands	1.2%	Feet	2.9%	*Percentage of Trunk:*	
Forearms	3.2%	Shanks	9.4%	Thorax	37.5%
Upper arms	5.4%	Thighs	20.4%	Abdomen and pelvis	48.8%
		Whole trunk (including head and neck)	57.4%	Head and neck	13.7%

Table 3-2 Specific Gravity (Dempster)

Hand	1.16	Arm	1.07	Whole trunk	1.03
Foot	1.10	Thigh	1.05	Head and neck[a]	1.11
Forearm	1.13	Abdomen and pelvis	1.01	Thorax[b]	.92
Shank	1.09				

[a] Neck-thorax separation — upper border of first rib and between C-7 and T-1.
[b] Thorax-abdomen separation — peritoneum from lower surface of diaphragm and between T-12 and L-1.

Table 3-3 Segment Lengths as Percentages of Total Height (Dempster)

Upper arm	16.9%	Thigh	23.4%	Hip to shoulder	30.4%
Forearm	15.9%	Shank	25.3%		

Dempster gave the locations of the centers of gravity in terms of both a percentage of segment length and a percentage of the hip joint to the top of the head distance for the trunk. The trunk data have been changed in this text's analysis so that the percentages can be taken from the commonly used length for motion analysis, hip joint to shoulder joint. In addition, it is necessary to calculate the radii of gyration (k) from Dempster's data on moments of inertia (I) by using the formula $k^2 = I/m$, where m = mass. The radii of gyration have also been determined about the distal ends by use of the parallel axis theorem, as follows:

$$mk_d^2 = I_{c.g.} + mb^2$$
$$-mk_p^2 = I_{c.g.} + ma^2$$
$$\overline{k_d^2 - k_p^2 = b^2 - a^2}$$

or

$$k_d^2 = k_p^2 + b^2 - a^2$$

where a = distance from the proximal (P) end to the center of gravity,
 b = distance from the distal (d) end to the center of gravity, and
 $I_{c.g.}$ = moment of interia about the center of gravity.

Anatomical Data

Table 3-4 The Location of the Segment Center of Gravity From the Proximal and Distal Joint (r_p, r_d) and the Radius of Gyration From Each End (k_p, k_d). (From Dempster, B-7)

Segment	r_p	r_d	k_p	k_d
Hand	50.6	49.4	58.7	57.7
Forearm	43.0	57.0	52.6	54.7
Upper arm	43.6	56.4	54.2	54.5
Foot	50.0	50.0	69.0	59.0
Shank	43.3	56.7	52.8	54.3
Thigh	43.3	56.7	54.0	55.3
Trunk (including head) (e = hip to shoulder)	66.0	34.0	83.0	50.7
Forearm and hand (elbow to ulnar styloid)	68.2	31.8	82.7	56.5
Whole upper limb (shoulder to ulnar styloid)	53.0	47.0	64.5	59.6
Foot and shank (knee to med. mall.)	60.6	39.4	73.5	57.2
Whole lower limb (hip to med. mall.)	44.7	55.3	56.0	55.0

Additional data have been contributed by Parks (B-17), who made a more detailed dissection of one trunk. He followed the same dissection planes as Dempster and in addition divided the abdomen and pelvis. After data were obtained on the trunk parts, additional dissections of the thorax and abdomen were performed by cutting horizontal cross sections of approximately one inch (Fig. 3-2). This determined a line of centers of gravity (Fig. 3-3) that was used in conjunction with the distribution of body mass relative to body height as determined by Dempster (Fig. 3-4). This produced a method that makes it possible to determine the trunk center of gravity during changing positions

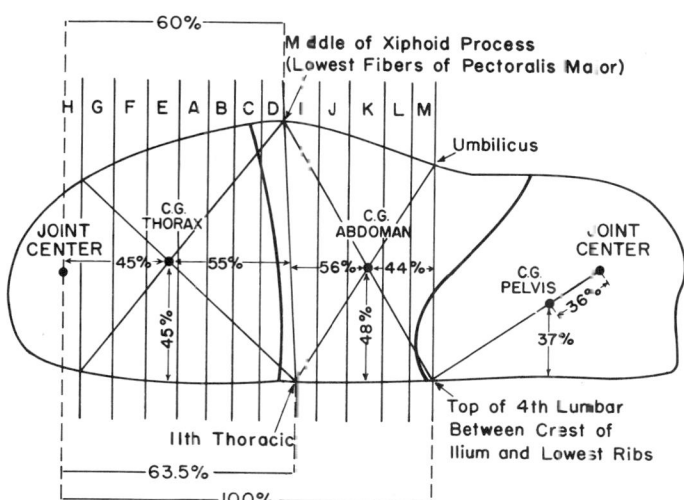

Fig. 3-2
Trunk segments—centers of gravity (Parks).

Anatomical Data

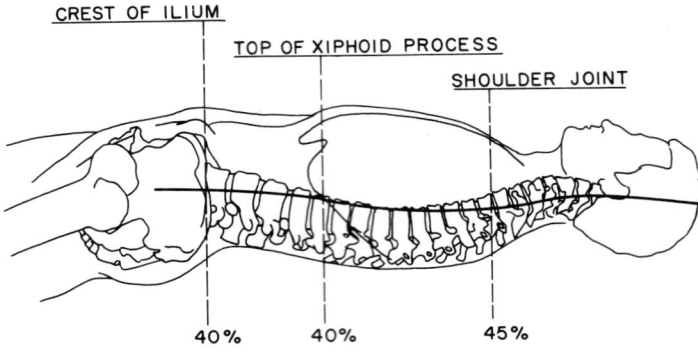

Fig. 3-3
Trunk-line of centers of gravity (Parks). Percentage equals distance upward from dorsal side to the center of gravity line.

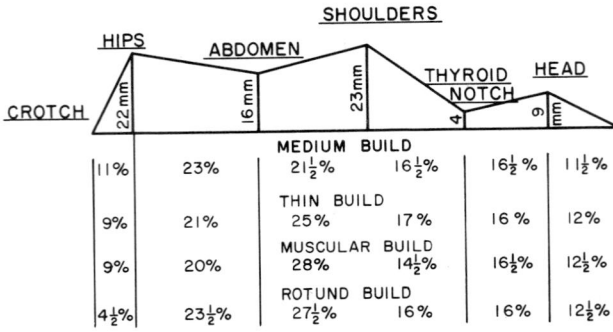

Fig. 3-4
Distribution of trunk mass (Dempster). Percentage equals the part of the total trunk length.

Fig. 3-5
Cardboard cutout–lead method to determine trunk center of gravity.

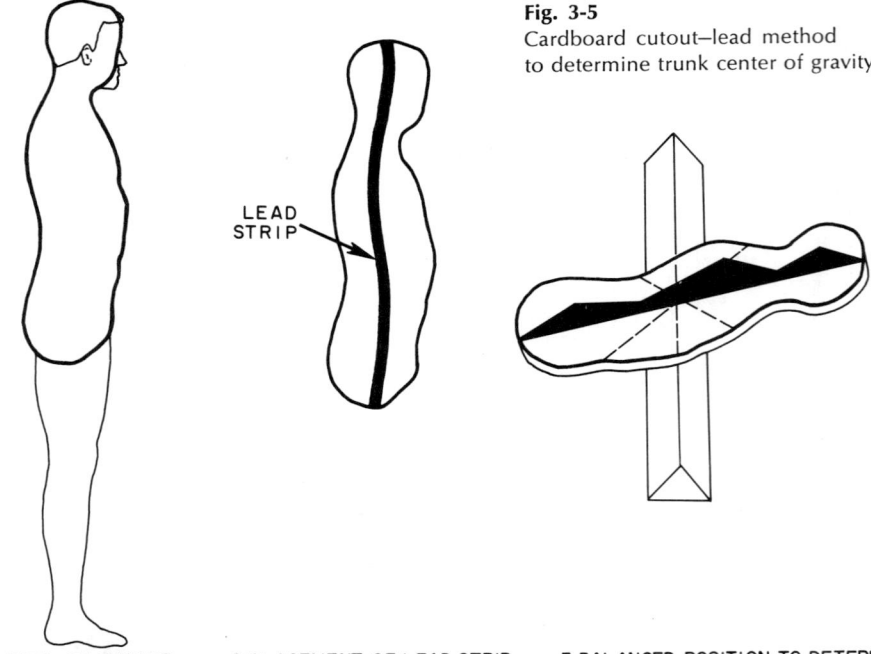

1. AREA OF CUTOUT
2. PLACEMENT OF LEAD STRIP
3. BALANCED POSITION TO DETERMINE CENTER OF GRAVITY

of extension and flexion when the trunk cannot be considered rigid. This is done by making a cardboard cut-out of the trunk for the position to be analyzed, drawing a line of the centers of gravity, and placing a lead strip to correspond to the distribution of body mass (Fig. 3-5). The lead strip must be cut in the proportion 9:4:23:16:22 at the proper points on the trunk (1, the widest part of the head; 2, the neck at the thyroid notch; 3, the shoulder joint; 4, the umbilicus; 5, the hip joint; (see Fig. 3-4). These significant points are given as a percentage of the total trunk height, so application may be made to any trunk size. The corresponding proportions for different body builds are: thin, 8:3:18:11:18; muscular, 9:4:23:15:23; rotund, 9:5:28:27:30. The center of gravity is determined by balancing the system on a knife edge in three positions (see Fig. 3-5).

An additional change in the trunk center of gravity occurs when the arms are raised and the shoulder mass is elevated. The shoulder mass was calculated to be 5.5% of the total trunk weight by Dempster, so a correction based on this figure has been estimated. It was calculated that a change of 10 cm in the hip joint to gleno-humeral joint would raise the total trunk center of gravity 1.8 cm. The accuracy of this calculation needs verification using a cadaver dissection. (This correction in the normal trunk center of gravity is also written into the computer program — see Appendix B).

It was necessary to obtain the trunk center of gravity and radius of gyration without the head and neck in order to analyze a motion where the head is considered as a separate segment (soccer heading). It was also necessary to obtain the radius of gyration of the whole trunk including the head when the pivotal point is the shoulder joint (for movements that have the hands as the fixed point, such as horizontal bar and parallel bar exercises in gymnastics). This was obtained by using the cardboard-lead cut-outs, and these data should also be verified by a cadaver dissection (Figs. 3-6 and 3-7).

Fig. 3-6
Center of gravity of the head and neck, and the remaining trunk (male).

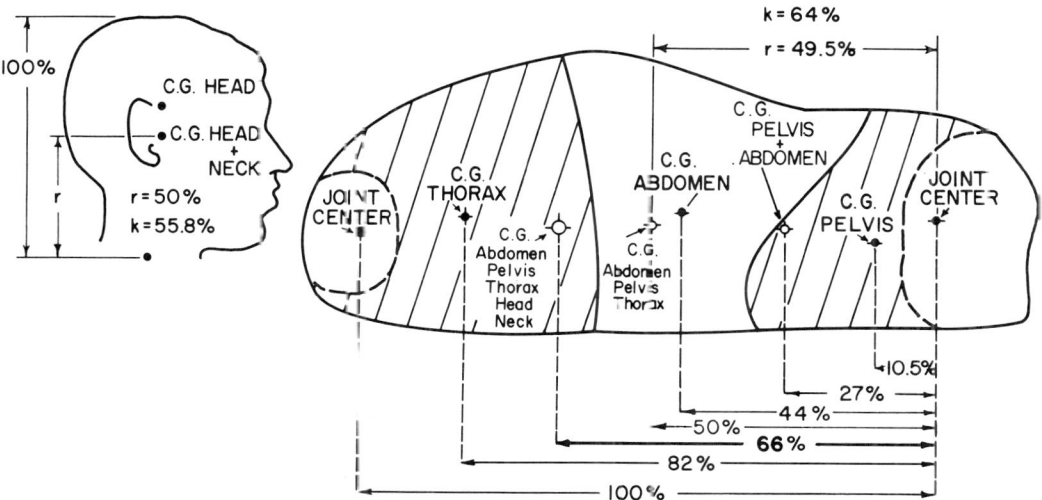

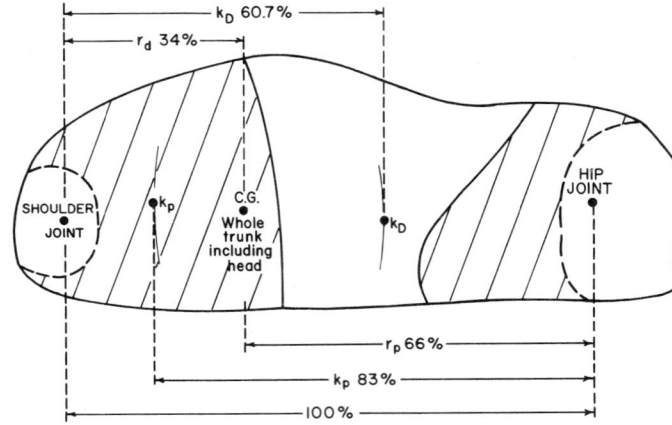

Fig. 3-7
Proximal and distal r and k of the whole trunk (including the head and neck).

Until more cadaver dissections are performed, the moment of inertia (I) and the radius of gyration (k) of the trunk moving about varying axes must be calculated using data from known shapes. The r and k for the whole trunk, including head and neck, as the trunk revolves about the hips (horizontal axis) are 66% and 83%, respectively, of the hip to shoulder distance. If the trunk is considered a cylinder, the moment of inertia would be (see Fig. 3-8):

$$I = \tfrac{1}{12}m(3R + 4L^2)$$

where m = mass. The R of the cylinder is compared to an r using body measurements for determining the moment of inertia when the L of the cylinder is made equal to the distance from the hip joint level to the top of the head. The r is determined by measuring the width and depth of the trunk at the level of the anterior superior iliac spine of the pelvis and calculated using the equation given in Fig. 3-8.

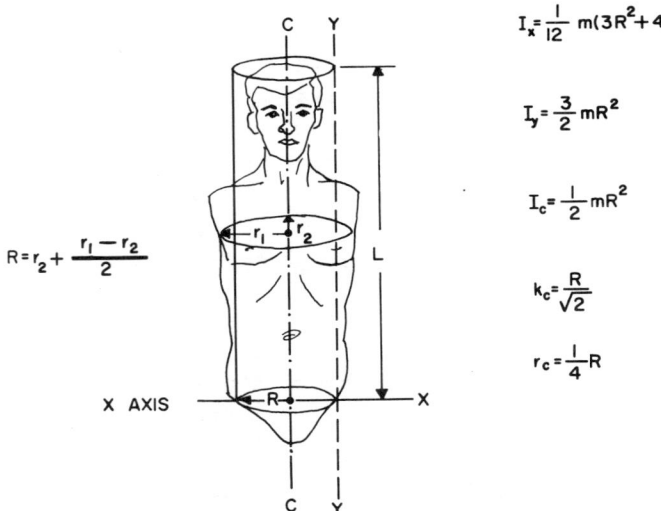

Fig. 3-8
Comparison of the trunk to a cylinder to aid in the determination of the radius of gyration about varying axes.

Anatomical Data

From the appropriate measurements of five college men the radius of gyration averaged 53.5% of the hip to shoulder distance (about the X axis), or within 0.5% of the cadaver data as determined by Dempster. Therefore, until more accurate data are obtained, the trunk may be considered to be a cylinder. If the trunk revolves about a longitudinal (Y) axis passing through one hip joint (kicking, soccer style), then $I = 3mr^2/2$ and $r = 50\%$ and $k = 61\%$ of the distance between the two hip joints. If rotation occurs about the central (C) axis, then $r = 25\%$ and $k = 35\%$.

By use of the specific gravity data and the locations of trunk centers of gravity determined by Dempster, segment weights of living men (35 college-age students) were determined using the water displacement method. The trunk segments were weighed by submerging to the plane of the saw cuts with the air expelled. The body segment weights as percentages of total body weight for living men are presented in Table 3-5. The segment lengths as percentages of height are presented in Table 3-6.

Table 3-5 Body Segment Percentages of Total Body Weight for Living Men ($N = 35$)

Hands	1.3	Thighs	21.0	*Percentages of Trunk*	
Forearms	3.8	Whole trunk		Head	12.8
Upper arms	6.6	(including		Neck	4.5
Feet	2.9	head and		Thorax	33.7
Shanks	9.0	neck)	55.4	Abdomen	24.2
				Pelvis	24.8

Table 3-6 Segment Length Percentages of Total Height for Living Men ($N = 35$)

Upper arm	17.2%	Thigh	23.2%	Hip to	
Forearm	15.7%	Shank	24.7%	shoulder	30.0%

Similar data have always been needed for women. The water displacement technique was used on 76 college-age women to obtain body segment percentages of total body weight, and on seven modern dancers. Kjeldsen (B-13) used the same technique on six women gymnasts and six selected nongymnasts whose hip widths were greater than shoulder widths. Kjeldsen then divided the trunk into three segments and used the water displacement technique to determine the percent of trunk weight of the head and neck, thorax, and abdomen-pelvis [using the section planes described by Dempster (Table 3-2)]. Table 3-7 summarizes body segment percentages of total body weight and trunk segment percentages of trunk weight. Table 3-8 presents segment length percentages of total height.

With this information, Kjeldsen determined the trunk center of gravity for women and calculated the proximal and distal radii of gyration. Figure 3-9a–d presents these data. (The percentages of limb lengths to obtain r's and k's for women must remain the same as for men until female dissections are performed.)

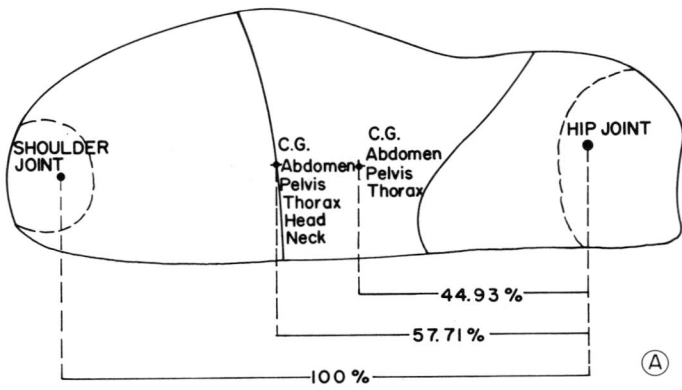

Fig. 3-9a
Woman gymnast—trunk center of gravity (Kjeldsen).

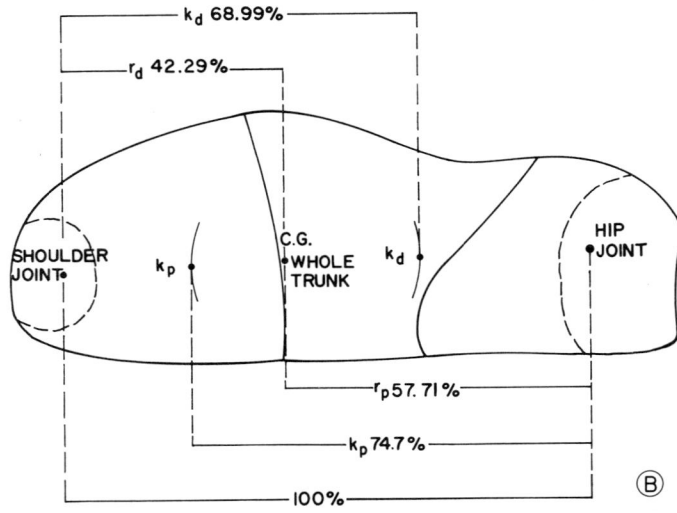

Fig. 3-9b
Woman gymnast—proximal and distal r and k of the whole trunk (Kjeldsen).

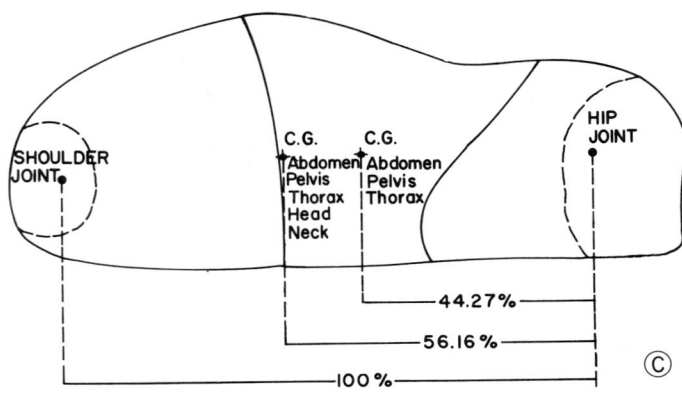

Fig. 3-9c
Woman—hip width greater than shoulder width, trunk center of gravity (Kjeldsen).

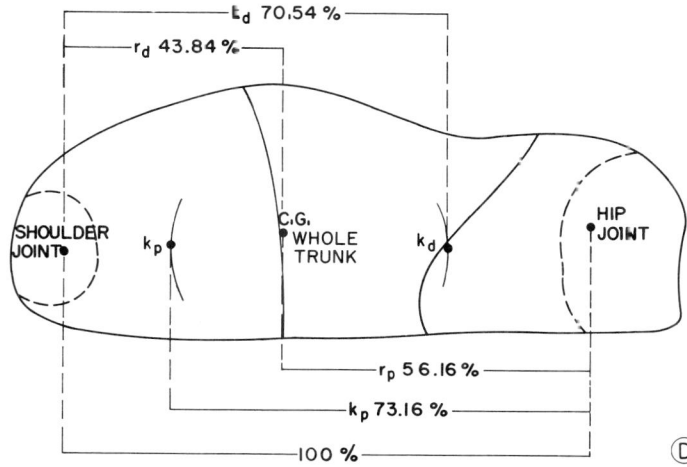

Fig. 3-9d
Woman—hip width greater than shoulder width, proximal and distal r and k of the whole trunk (Kjeldsen).

Table 3-7 Body Segment Percentages of Total Body Weight for Living Women ($N = 76$)

	College Age ($N = 76$)	Dancers ($N = 7$)	Gymnasts ($N = 6$) (Kjeldsen)
Hands	1.0	1.0	1.03
Forearms	3.1	2.9	3.21
Upper arms	6.0	5.8	5.49
Feet	2.4	2.7	2.48
Shanks	10.5	11.0	10.98
Thighs	23.0	24.3	16.52
Whole trunk (including head and neck)	54.0	52.3	60.24

Percentage of Trunk (Kjeldsen)

	Nongymnasts ($N = 6$)	Gymnasts ($N = 6$)
Head and neck	14.33	15.61
Thorax	27.44	27.52
Abdomen and pelvis	58.89	56.74

Table 3-8 Segment Length Percentages of Total Height for Living Women

	College Age ($N = 49$)	Dancers ($N = 7$)	Gymnasts ($N = 6$) (Kjeldsen)
Upper arm	17.2	16.5	19.3
Forearm	14.7	16.2	16.6
Thigh	24.8	25.1	24.7
Shank	24.4	25.2	25.6
Hip to shoulder	29.6	30.1	30.0

With the presentation of the preceding anatomical data, it is now possible to continue the procedures for analyzing motion. The segment lengths and angles may be measured from the tracing, and segment weights, centers of gravity, and radii of gyration are available from the anatomical data given above.

chapter 4

THE FREE BODY DIAGRAM FOR A ONE-SEGMENT MOTION

Whether the record of a body motion is obtained using motion pictures, strobe pictures, or the electrogoniometer, the next step is to determine the angular velocities and accelerations at instantaneous positions. This is done by measuring the angle of the segment from the right horizontal, preferably using a drafting machine. Figure 4-1 shows the standard mathematical coordinate system with the plus vertical direction upward, the plus horizontal direction to the right, and the plus direction of angular motion counterclockwise. The calculation of the forces due to motion ($mr\alpha$ and $mr\omega^2$) and the drawing of a free body diagram follow the determination of angular velocities and accelerations.

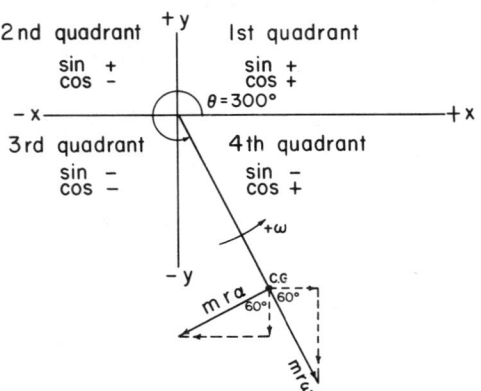

Fig. 4-1
Coordinate system and force directions as drawn using a free body diagram.

DISPLACEMENT CURVE

The Free Body Diagram for a One-Segment Motion

Drawing the displacement curve of a segment measured from the right horizontal is called the *absolute motion* method, because the segment's change of position is relative to the earth. A displacement curve is obtained by recording the number of degrees a segment has moved from the starting point (Fig. 4-2a). Table 4-1 presents the positions of the segment measured from the right horizontal, and the corresponding displacements from the starting point.

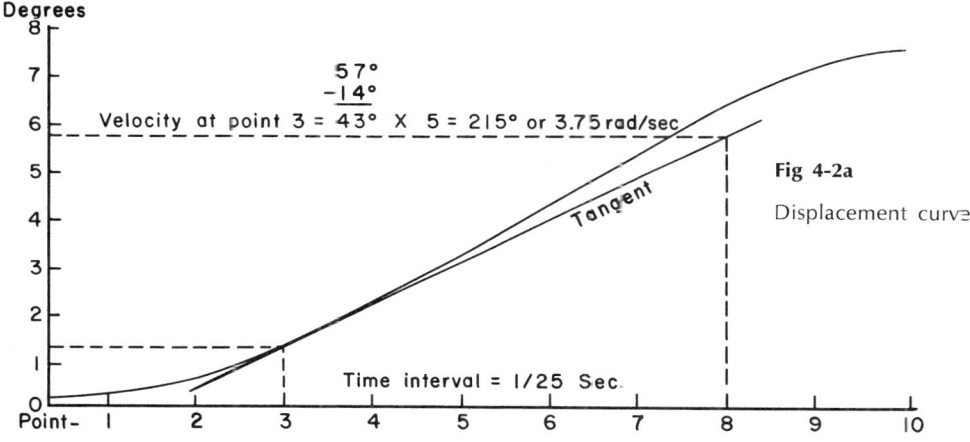

Fig 4-2a

Displacement curve.

Table 4-1 Data for Obtaining Body Segment Displacement

Position	Segment Position	Displacement
0	13°	0
1	11°	2°
2	6°	7°
3	1°	12°
4	351°	22°
5	341°	32°
6	331°	42°
7	321°	52°
8	310°	63°
9	302°	71°

If a change is measured relative to the adjoining body segment it is termed *relative motion*. After measuring angles for displacement curves, it should be clear that the first segment for a whole body motion is measured from a fixed point so the absolute and relative motions are thus the same; however, the second and succeeding segments differ due to the moving pivot point.

The Free Body Diagram for a One-Segment Motion

VELOCITY CURVE

The velocity curve is obtained from the displacement curve. A tangent drawn to the displacement curve at each position determines the rate of change per unit time at that instantaneous position. The measurement of the slope of the tangent produces the instantaneous velocity (first derivative). Position 3 in Fig. 4-2a shows how to obtain the velocity in degrees per second by measuring the change during 5 time intervals and multiplying by 5 to obtain the change per second. (This can also be done by measuring only one interval and multiplying by 25, but any measurement error would also be multiplied by 25). This process must be followed for each position and the velocity changed to the standard radians-per-second using a book of standard tables (1 radian = 57.3 degrees). The instantaneous velocities are then plotted to obtain the curve in Fig. 4-2b.

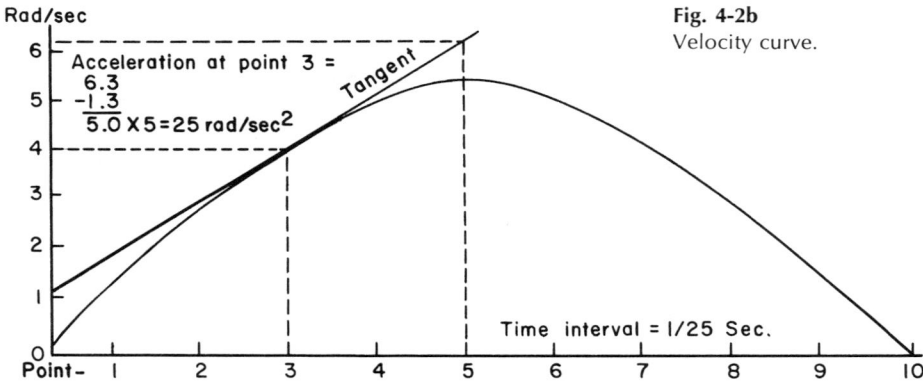

Fig. 4-2b
Velocity curve.

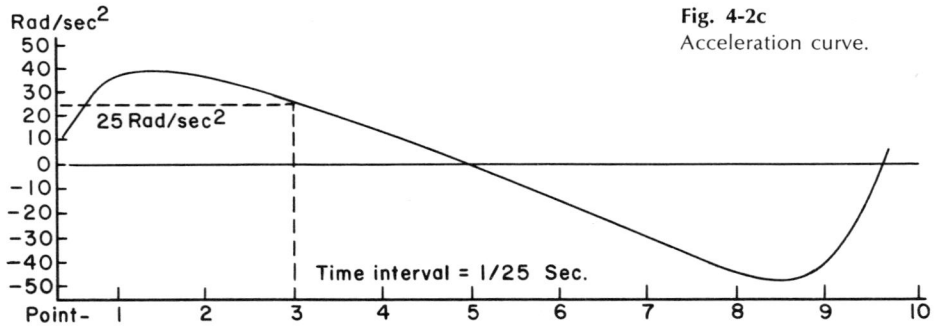

Fig. 4-2c
Acceleration curve.

ACCELERATION CURVE

A tangent drawn at each position on the velocity curve determines the velocity change per unit time (second derivative). The slope of each line is determined as shown for position 3 in Fig. 4-2b. The change in velocity was determined per second to obtain acceleration in radians per second per second

31

The Free Body Diagram for a One-Segment Motion

(rad/sec²). The plotted acceleration curve is shown in Fig. 4-2c. When drawing the curves, the usual procedure of smoothing out the plotted points with a French curve is employed. Every student should perform this hand plotting method to fully understand how instantaneous velocities and accelerations are obtained. This time-consuming process should not be needed again as the computer program (see Appendix B) calculates all velocities and accelerations, and smooths the curves according to the closeness of the curve fit desired.

STATIC EQUILIBRIUM

All the information needed is now available for calculating the sum of the segment forces and joint moments of force. Before attempting the actual analysis, however, the student must have a thorough understanding of the free body diagram. This requires a review of problems dealing with vectors, torque, and functions of a right triangle. Selected problems in these areas that will aid in motion analysis are presented in Appendix A. It is also necessary to determine the forces due to motion and to relate the free body diagram to the human body (in terms of muscle function and joint forces). This, in addition, requires a knowledge of the radius of gyration and centers of percussion. These essentials are covered briefly here before we attempt the actual calculations of forces.

Special references that will prove helpful are C-11, C-22, C-24, C-25, and C-40. The student should also work the problems in Appendix A.

A problem that involves all the mathematics required for an analysis of a static position related to the body is presented in Fig. 4-3. The wood suspended by a wire has its end against the building and is free to move except for friction. This system can be compared to the body by considering the wood as the forearm, the wire as the muscle, and the junction of the building and wood as the elbow joint (Fig. 4-3a).

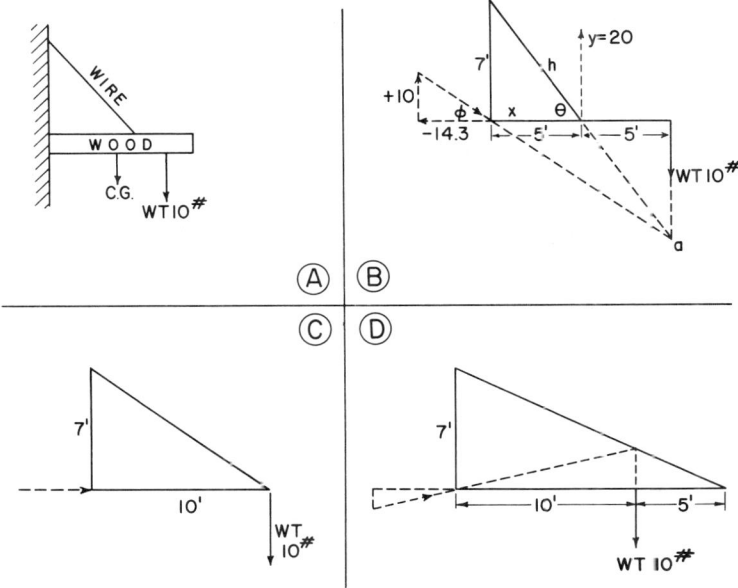

Fig. 4-3a, b, c, d
Static analysis of forces and force directions. (Wood-wire problem.)

If a weight is placed outside the wire attachment, as in Fig. 4-3b (wood weight disregarded), what is the direction and magnitude of the building-on-wood force? What is the force on the wire? Will the wood slip if the coefficient of friction is .6?

$$F \cdot D = F \cdot D \qquad \tan \theta = \frac{7}{5} = 1.4$$

$$\theta = 54°28'$$

$$10 \cdot 10 = 5y \qquad \sin \theta = \frac{20}{h}$$

$$h = 24.5 \text{ lb}$$

$$y = 20 \text{ lb} \qquad \tan \theta = \frac{20}{x}$$

$$x = 14.3 \text{ lb}$$

The resultant sum of the y forces is $20 - 10 = +10$ lb, and the resultant x force is -14.3 lb. Tan $\phi = 10/14.3 = 7$ and $\phi = 35°$.

The coefficient of friction (μ) is obtained by dividing the force it takes to move an object along a table top by the force perpendicular to the table top (commonly written: $\mu = F/N$). N is used rather than weight because N is not equal to the weight when the table is inclined. For the above problem $F = 10$ lb and $N = 14.3$ lb, so $\mu = .7$. Therefore, with the coefficient of friction between the building and wood less than .7, the wood would slip upward.

For the arrangement of wood, wire, and weight shown in Fig. 4-3c, $y = 0$ and $x = 14.3$. No slippage would occur regardless of the amount of weight suspended.

For the arrangement in Fig. 4-3d, $y = -3.34$ and $x = -14.3$. Friction less than .234 would cause slippage downward.

An additional method for obtaining the direction of the force at the building and wood juncture is to extend the wire and weight vectors until they meet. A line drawn from this point (point a in Fig. 4-3b) through the contact point of the building and wood will correspond to the 35° calculated above. This is possible because the *lines of action of three nonparallel forces intersect at a common point of a system in equilibrium.* This method may be used to obtain the force direction, without any calculations if magnitude is not desired. Calculate the forces with this method using Figs. 4-3c and d. Which diagram in Fig. 4-4 is correct? (Answer: D.)

If this force direction actually occurred at an elbow joint, the ligaments would prevent bone slippage, but more probably other muscles would contract to change the direction of the upward force. The role of joint stabilization fulfilled by continuously changing muscle contractions due to motion needs further study. In all probability, the muscles are continuously adjusting to the changing forces by attempting to maintain the joint bone-on-bone force as a compressive force in the line of direction least likely to allow articular displacement.

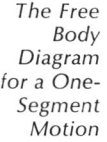

The Free Body Diagram for a One-Segment Motion

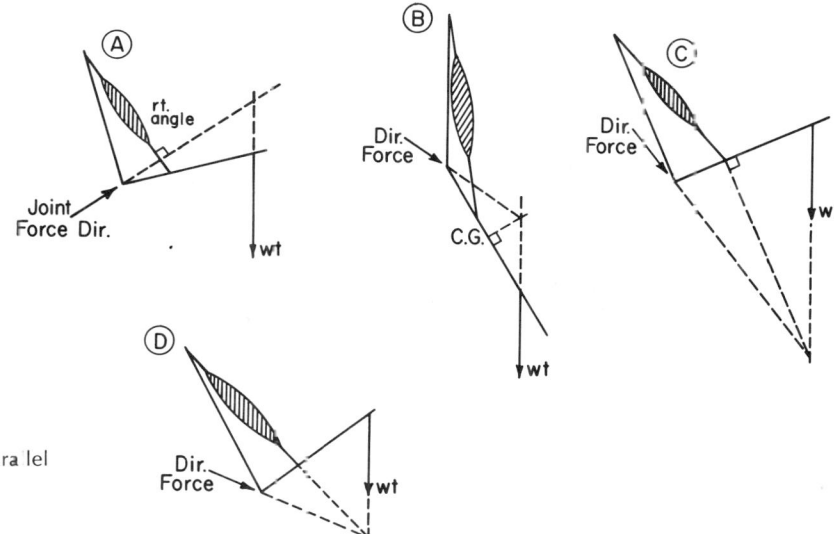

Fig. 4-4
The intersection of three nonparallel forces to determine the joint force direction. (Problem.)

FREE BODY DIAGRAM

The free body diagram is the isolated drawing of a body segment showing all the forces (vectorally) acting on it. This aids in calculating the segment forces and the joint moments of force. To draw a free body diagram of a segment in motion, all inertial forces must be drawn in the opposite direction to show the system in equilibrium. This was first realized by the French mathematician D'Alembert and is known as *D'Alembert's principle*. (All inertial forces in a free body diagram are treated in terms of equal and opposing forces, thus reducing a problem in dynamics to a problem of statics. All external forces and inertial forces form a system in equilibrium). Dempster's (D-5) work on the free body diagram best supplements the information presented here. (A glossary of terms and symbols for drawing a free body diagram is found at the end of Chapter 4.)

An elevator problem illustrates D'Alembert's principle. Everyone has felt the changing forces at the start and finish of an elevator ride. When the elevator starts upward you are pressed against the floor and you would weigh more if you were standing on a scale. Likewise you would weigh less when starting downward.

Problem 1 An elevator moving downward accelerates at 3 ft/sec/sec. How much would a 193 lb person weigh if he were standing on a scale? See Fig. 4-5.

Fig. 4-5
Inertial force drawn opposite to the direction of movement. (Elevator problem.)

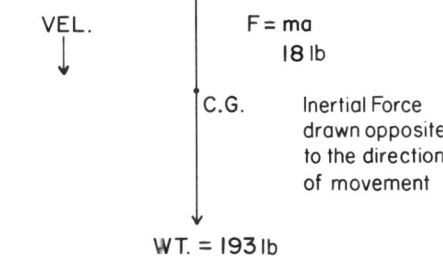

The Free Body Diagram for a One-Segment Motion

$$F = ma$$

$$F = \frac{193}{32.2} \times 3$$

$$F = 6 \times 3 = 18 \text{ lb}$$

Therefore, the scale reading equals 175 lb.

Problem 2 An elevator weighs 5000 lb. The cable holding it will break at 6000 lb. How fast would the elevator have to accelerate upward before the cable would snap if it carried three people each weighing 200 lb? (Answer: 2.3 ft/sec/sec.)

FORCES DUE TO MOTION

Before continuing with the study of the free body diagram the forces due to motion must be introduced, so that they can be properly illustrated on the free body diagram of an isolated segment. Newton's second law is the basis for most of the mechanics used to analyze body motion ($F = ma$), and the forces due to rotational motion based on this law are as follows.

The velocity of a stone on the end of a string rotating at a constant velocity is:

$$V = \frac{2\pi r}{t} \quad \text{or} \quad \frac{\text{circumference}}{\text{sec/revolution}}$$

The tangential velocity is at a right angle to a radius drawn to any instantaneous position specified, because the tangent represents the direction of the motion of the stone at that point (Fig. 4-6). Since $F = ma$, the tangential force F therefore equals $mr\alpha$, where α is angular acceleration.

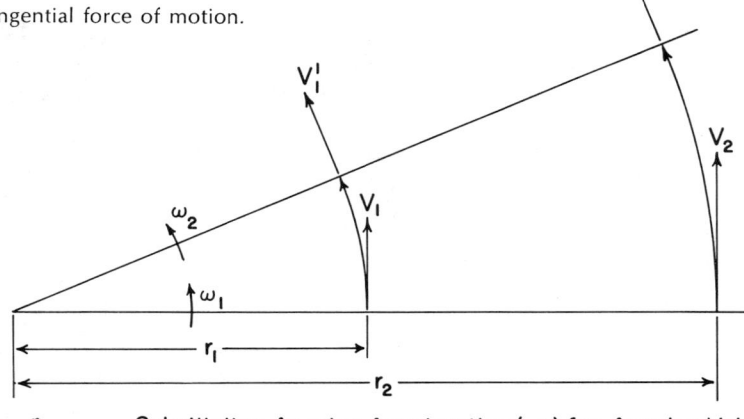

Fig. 4-6
Determination of the tangential force of motion.

$V_1 = \omega_1 r_1$ $V_1' = \omega_2 r_1$

$V_2 = \omega_1 r_2$ $V_2' = \omega_2 r_2$

Substituting Angular Acceleration (α) for Angular Velocity (ω) Linear Acceleration (a) becomes: $a_1 = \alpha_1 r_1$ $a_1' = \alpha_2 r_1$

$a_2 = \alpha_2 r_2$ $a_2' = \alpha_2 r_2$

Ans. $F = ma = mr\alpha$

The Free Body Diagram for a One-Segment Motion

However, in Fig. 4-7a there is a directional change in the velocity from 1 to 2, 2 to 3, etc., so an additional force exists. If the tangential velocity vectors are drawn to form another circle (Fig. 4-7b), the radius of the circle would be equal to the velocity, and the motion at each instantaneous point (tangent to V) would equal the change in velocity $a = 2\pi v/t$. If this acceleration vector is placed back into the first circle (Fig. 4-7a), its direction is normal to the radius toward the center.

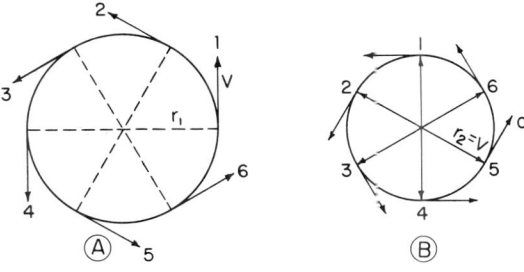

Fig. 4-7a, b, c
Determination of the normal force of motion.

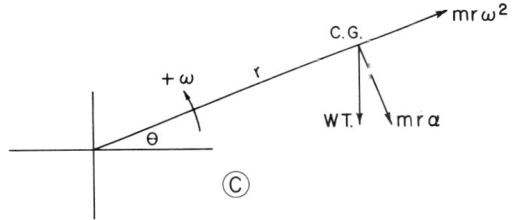

We now have enough basic equations so that we can make substitutions to arrive at an equation for centripetal force:

$$V = \frac{2\pi r}{t} \qquad a = \frac{2\pi v}{t}$$

$$t = \frac{2\pi r}{V} \qquad t = \frac{2\pi v}{a}$$

therefore,

$$\frac{2\pi r}{V} = \frac{2\pi v}{a} \qquad \text{solving for } a \qquad a = \frac{v^2}{r}$$

$$F = ma = \frac{mv^2}{r} \qquad v = r\omega$$

$$F = \frac{mr^2\omega^2}{r} = mr\omega^2 = \text{centripetal force (normal to } r \text{ toward center)}$$

Figure 4-7c illustrates these inertial forces drawn on a free body diagram, with the weight of the segment always drawn downward. The tangential force ($mr\alpha$) is drawn perpendicular to the segment in the direction opposite to the movement, and the normal force ($mr\omega^2$) is drawn outward, parallel to the segment. This direction of the tangential force is correct only if α is positive.

The Free Body Diagram for a One-Segment Motion

If α is negative, the force changes to the opposite direction, just as the force of a decelerating elevator is the opposite of one accelerating in the same direction. Therefore, while $mr\omega^2$ is always drawn outward from the pivot point regardless of the direction of the movement, $mr\alpha$ is drawn opposite to the direction of the movement when α is plus, and in the same direction as the movement when α is negative. Given these facts, which part of Fig. 4-8 a–b is correct? (Answer: B, A.)

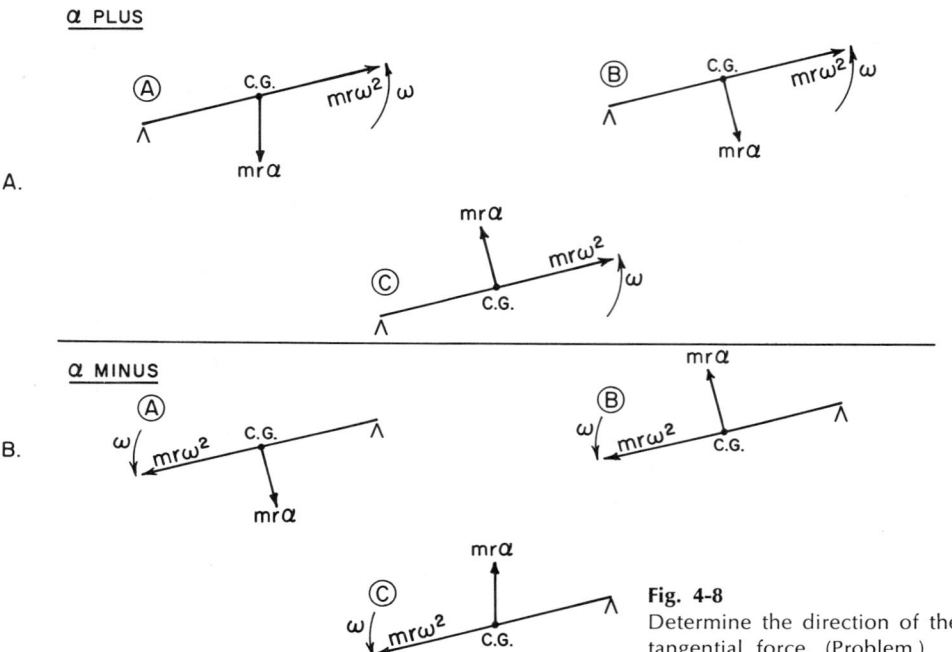

Fig. 4-8
Determine the direction of the tangential force. (Problem.)

The free body diagram for a one-segment motion is still not complete, because the segment is also influenced by the bone-on-bone force due to the adjoining segment (F_u), and by all the contracting muscles that are attached to the segment (F_m). Figure 4-9 includes these forces and completes the free body diagram for one segment in motion.

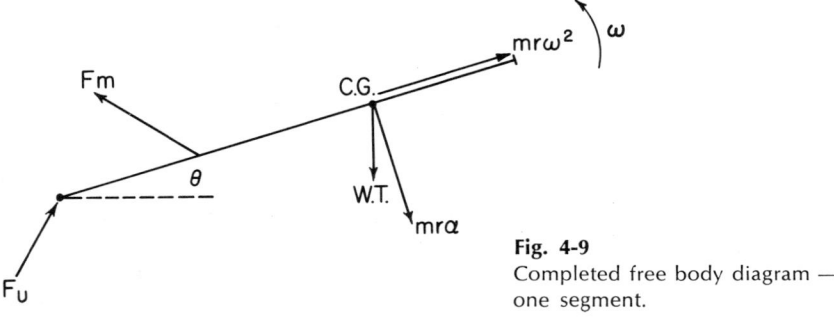

Fig. 4-9
Completed free body diagram — one segment.

The Free Body Diagram for a One-Segment Motion

This greatly complicates an analysis because the number of muscles involved, the extent of their contractions, and the actual distance of their insertions from the joint center are not known. Due to these unknowns, the actual direction of the bone-on-bone joint forces (F_u) cannot be determined. Valuable information results from the analysis, however, since the sum of all muscle forces times the sum of their resultant insertion distances from the joint center (M_o) equals the moments of force that must exactly balance the moments of force from the calculated external and inertial forces:

$$M_o = \text{wt}\,(\cos\theta)r + m\alpha \times r + mr\omega^2 \times 0$$

or

$$M_o - \text{wt}\,(\cos\theta)r - mr\alpha \times r = 0$$

(A later discussion of radius of gyration and moment of inertia will show that $I\alpha$ is used rather than $mr\alpha \times r$).

The determination of the moments of force represent the dominant muscle action producing the segment motion. The actual extent of the contraction is not known, but it changes as the speed of movement changes. If a forearm and hand were moved slowly, with no added weight in the hand, the calculated moments should be very near the actual moments of the combined forearm flexors. However, if the same motion were repeated rapidly the moments representing muscle action would be much greater than the calculations indicated. Actually, the flexing moments would be greater because many muscles surrounding the joint contract, performing their function of joint stabilization. The extent of the dominance of the flexors over the muscles that are noncontributors to the flexing action is not known. Dempster and Finerty's study (J-17) shows that these muscles contract one-third as much as the prime movers for hand flexion, but a great deal more must be studied at varied speeds of motion. For the present, we must use the term *dominant muscle action* to interpret body motion relative to the external and inertial forces of motion.

To further illustrate the complexity of determining the actual bone-on-bone joint forces Fig. 4-10 shows the procedure that could be used if the muscles contracting could be isolated. Figure 4-10a shows the resultant force vector, due to the weight of the forearm and hand, and the forces due to motion ($mr\alpha$ and $mr\omega^2$) intersecting with the line of action of the biceps (point *a*). A line drawn from point *a* to the joint center produces the actual joint force direction if only the biceps were involved.

Figure 4-10b extends this procedure to two muscles. If the same movement were performed with only these two muscles contracting, an additional joint force, due to the line of pull of the triceps (vector **a**), would result (vector **b**). The actual bone-on-bone joint force would then be the resultant of vectors **b** and **c**. If ten muscles were involved in a given joint movement, this procedure would be necessary for each muscle. The resultant of all the directional forces could not be determined, however, because the magnitude of each vector cannot be ascertained. This treatment of isolating muscles is presented only to emphasize the fact that the field of kinesiology is a long way from determining the actual direction and magnitude of joint forces during motion, and that only generalizations can be made at present.

This type of isolated analysis may be worthwhile, however, if the electromyograph is used simultaneously. Two important body functions emerge when

The Free Body Diagram for a One-Segment Motion

the calculated forces are compared to EMG readings of a one-segment motion.

1. *The total joint muscle action adjusts constantly during a motion, attempting to maintain a stable joint* (to maintain a bone-on-bone direction which has one bone pushing into the socket of the other; compressive rather than shearing).
2. *The muscles that have their long axes closest to the resultant force vector direction are called on to the greatest extent.*

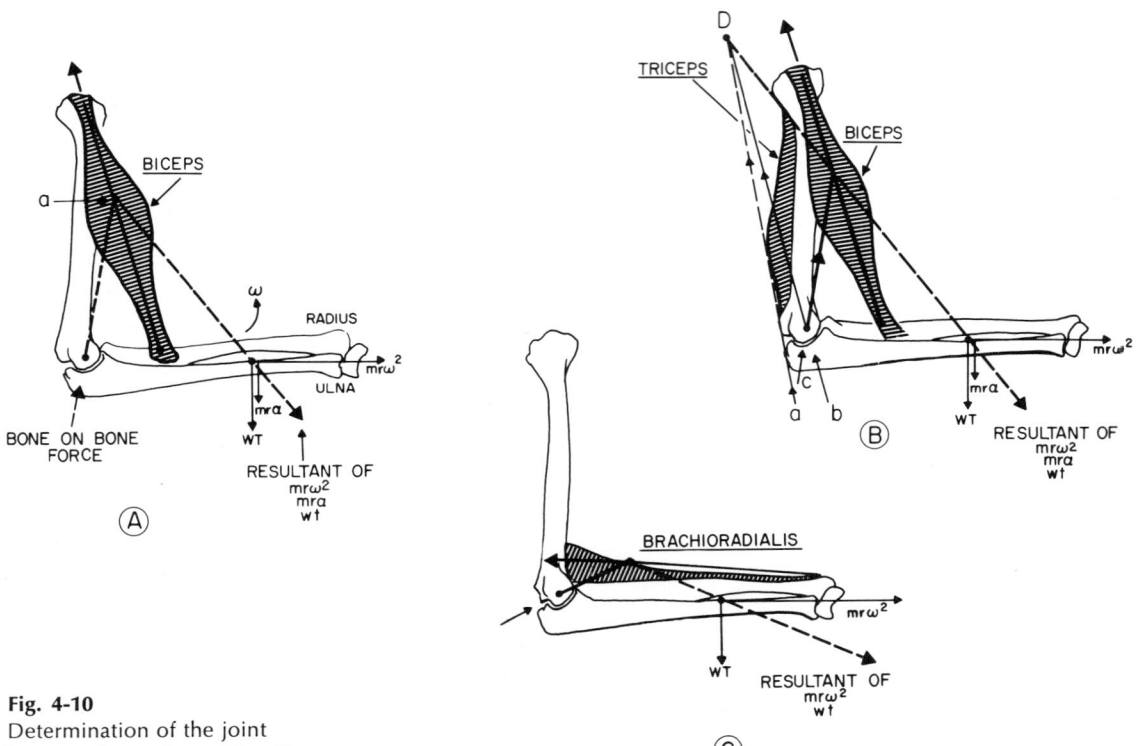

Fig. 4-10 Determination of the joint bone-on-bone force direction.

Figure 4-10c illustrates how the brachioradialis is in a favorable position and is called on more and more as the resultant force vector approaches the longitudinal axis of the forearm. EMG readings show the increase in the contraction of the brachioradialis as the angular velocity increases and reaches its maximum, while the acceleration reaches its minimum.

Figure 4-10 also aids in clarifying the terms *spurt* and *shunt muscles* described thoroughly by Basmajian (A-2, A-3). A spurt muscle (Fig. 4-10a) is in a position to best produce speed of movement and a shunt muscle is in a position to aid in joint stabilization (Fig. 4-10c). Because of the selectivity of the muscles used as the force directions change, it becomes possible to predict EMG readings (generally rather than specifically) from the calculated data of a force analysis.

The Free Body Diagram for a One-Segment Motion

Past kinesiology courses have traditionally listed the muscles that contract as indicated by the direction of the body segment movement. This list was generally limited to those muscles that started the segment motion, but failed to consider those that stopped the motion. Because of the constantly changing inertial forces and the difficulty of determining the changing role of each isolated muscle, it is unproductive to list individual muscles as a means of analyzing motion. Unless there is a need for obtaining data on a specific muscle, motion is best interpreted by referring to the dominant group of muscles for a given instantaneous position, which is obtained from the joint moments of force.

The general pattern of a one-segment motion is to have the maximum acceleration occur shortly after the start of the motion, the maximum velocity near the middle of the motion, and the maximum deceleration near the end of the motion. Even for such a simple motion, the eye cannot discern where the deceleration begins and thus where the change in dominant muscle action occurs. Previously, all the flexors of the gleno-humeral joint would be listed for a simple arm raising motion. However, the magnitude of the dominant flexors and the dominant extensors may be stated if the joint moments are calculated. Additional muscles are called into action to varying degrees when the arm raising motion is performed rapidly. A fast, one-arm-only motion stopped abruptly would use not just all shoulder flexors and extensors to varying degrees — joint stabilization would also to some extent call upon every other muscle surrounding the joint. In addition, the entire body remains rigid as a support to the arm movement, so trunk rotation and leg stabilization occur. Therefore, if the method of listing muscles is used, almost every muscle in the body could be listed for a simple fast arm raise. By using a moment of force analysis the dominant muscle action at all joints would indicate the extent of the muscle action occurring.

Unobservable inertial force changes that occur during a rapid motion are illustrated in Fig. 4-11. The acceleration is positive and large and the velocity

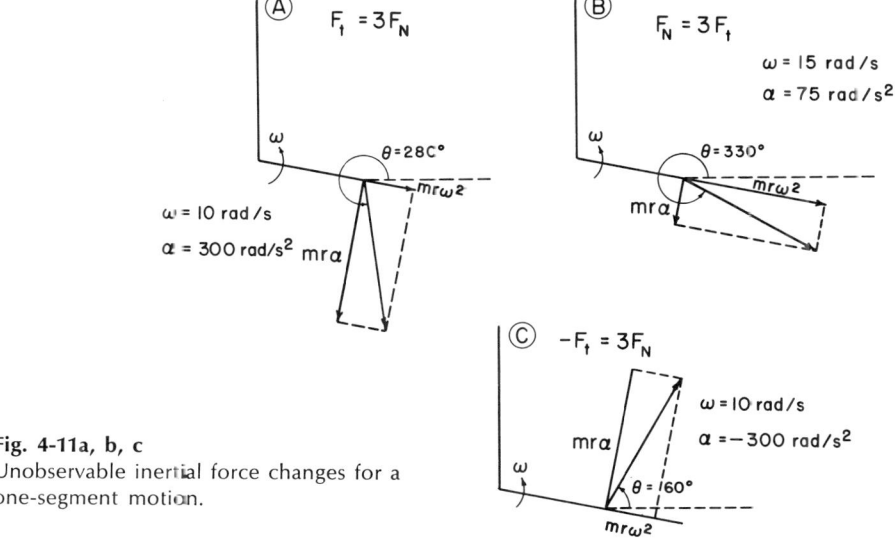

Fig. 4-11a, b, c
Unobservable inertial force changes for a one-segment motion.

small in Fig. 4-11a, thus producing a downward force almost perpendicular to the segment. The acceleration is small and the velocity large in Fig. 4-11b, thus producing a force approaching the longitudinal axis of the segment. The deceleration is large and the velocity small in Fig. 4-11c, so the force is almost perpendicular to the segment and upward. The magnitude of the moments of force indicates the extent of the muscle contraction, and it is evident that the dominant muscles are the flexors in Fig. 4-11a, the stabilizers in Fig. 4-11b, and the extensors in Fig. 4-11c. Because the eye cannot perceive the change from acceleration to deceleration, nor calculate the effect of gravity relative to the speed of motion, it is also evident that muscle action cannot be determined visually for a simple segment motion except in a very general way.

This points out the misuse of the term "antagonist." *Antagonistic muscles* are contracting muscles opposing the movement of the segment or opposing the prime movers. Muscles will not oppose a desired motion unless the movement is fast enough to require joint stabilization. In its role as a stabilizer, a muscle could also be antagonistic to the desired motion. If, however, these same muscles contract, thus fulfilling the role of deceleration of the movement, they are no longer antagonists but *prime stoppers*. This distinction is necessary to prevent the indiscriminate use of the word "antagonist."

APPLICATION OF THE FREE BODY DIAGRAM

If a person were standing on a scale, the scale reading would change with changes in inertial force magnitude and direction produced by the person's motions. Therefore, the stage in a motion shown in Fig. 4-11a would produce a scale reading heavier than body weight, that in Fig. 4-11b would show very little change, and the scale reading would be less than body weight for the point in Fig. 4-11c. A normal household scale cannot record these changes, but a force plate is designed to record the instantaneous changes of the total force both in the vertical and horizontal directions (sum of forces at the feet pressing against the floor plate). This affords a means for checking the accuracy of a mathematical analysis by comparing calculated data with force plate readings. It also aids in understanding the concept of the sum of forces, which is often confused with moments of force. No matter what body position is assumed, the force plate will record the body's weight when it is standing still. This is the sum of the weights of all body segments without regard to the distance of the force from the joint center. However, moments of force do vary with the distance of the force from the joint center. (Moments equal the perpendicular component of the weight to the segment, times the distance to the joint center.)

It can now be shown that the proper timing of body extremities (deceleration) aids a jumping action by reducing the vertical force (F_y) at take-off. This is done by attaining maximum deceleration of the arms just before the take-off in a standing broad jump so that the forceful straightening of the legs has less weight to lift (see Fig. 4-12). The same action occurs during the take-off of a high jumper or a broad jumper (see Fig. 4-13). The kicking leg and both arms should have a perfectly timed deceleration, so that the extension of the take-off leg is done against the smallest force possible. This arm lift and timed deceleration occurs in ski jumping when the arms go backwards and come to an abrupt stop. The jump out of a forward roll to a handstand shows the vigorous assistance of the arms in getting the body up and around (see Fig. 4-14).

The Free Body Diagram for a One-Segment Motion

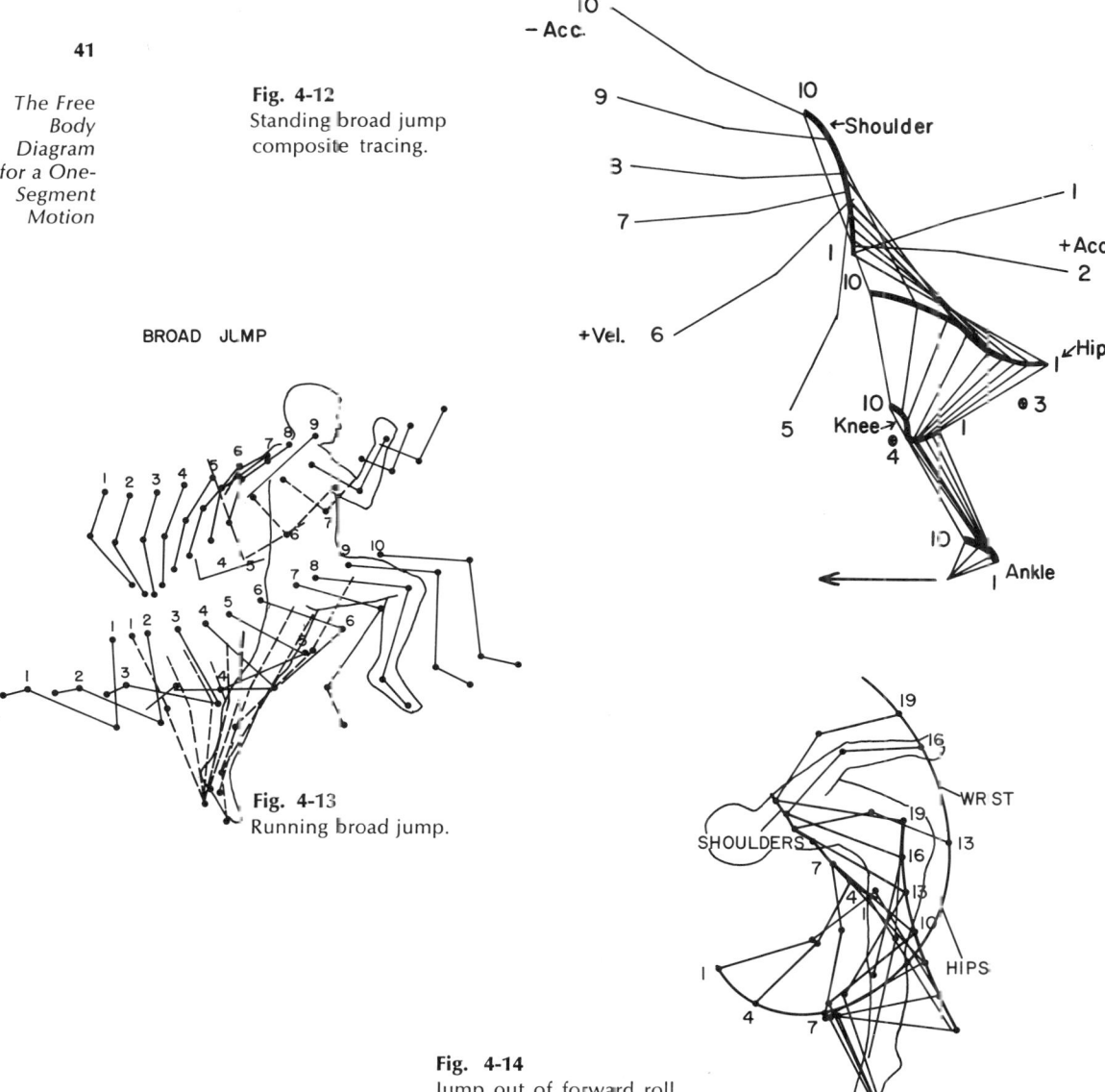

Fig. 4-12 Standing broad jump composite tracing.

Fig. 4-13 Running broad jump.

Fig. 4-14 Jump out of forward roll to handstand.

The jump of a basketball player going for a rebound or jump ball exhibits the same timing, but it is less obvious due to the continued reaching after take-off.

An additional example illustrating the importance of the arms is shown in Fig. 4-15. The arms are circled down and outward (Fig. 4-15a), with the acceleration of the arms reducing the force on the feet during the deep knee bend that follows landing. This movement ends with the arms forward (Fig. 4-15b), and the attempt to jump upward immediately after landing is difficult because the arms cannot aid the jump. Normally, a person starts with the arms back to attain a large velocity so that the deceleration will be significant and

42

The Free Body Diagram for a One-Segment Motion

aid the jump. Many students mistakenly believe that the arm movement helps a jump because of the momentum (mv) attained by the speed of movement of a segment, and that this momentum pulls the rest of the body along. Explaining the interactions of body parts in terms of inertial forces and their directions is the first step in understanding the importance of timing of the deceleration of one body part to aid the movement of the next body segment.

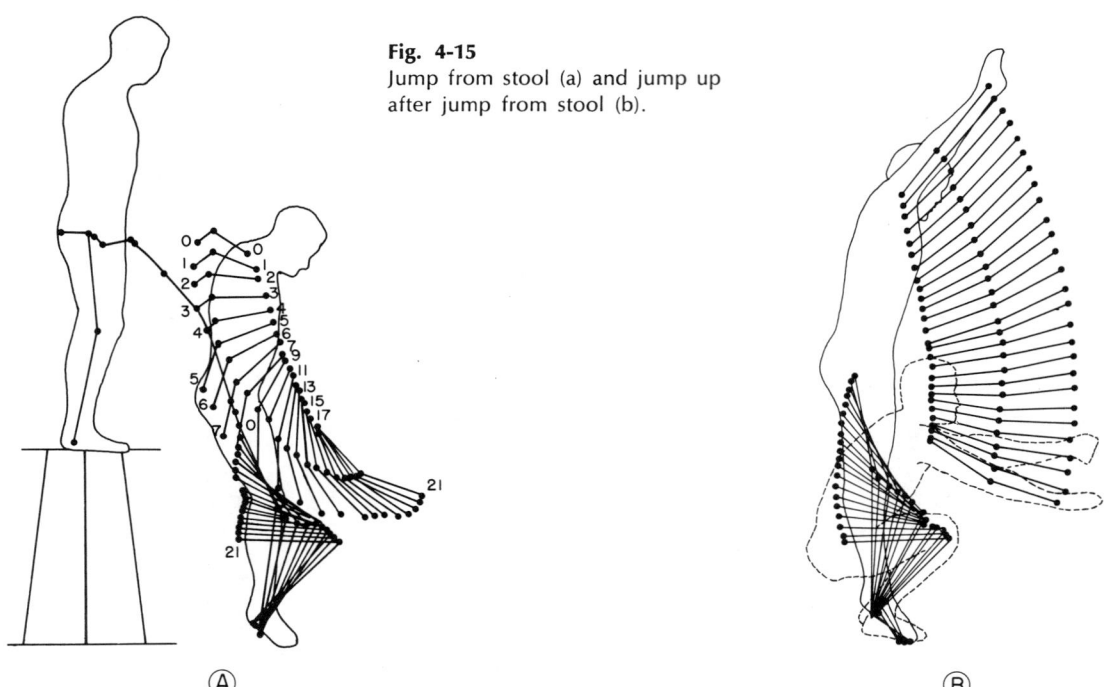

Fig. 4-15
Jump from stool (a) and jump up after jump from stool (b).

FLIGHT

After the feet leave the ground in the jumps illustrated above, the total body center of gravity follows the path of a parabola. (The ski jumper is an exception, where a problem in aerodynamics is introduced.) Knowing the path of a body in flight is important in sports because it is the only means of determining the angle of take-off (as in the broad jump), and it makes it possible to calculate the effect of air resistance on flight (Fig. 4-16). This parabolic path cannot be altered except by an outside force, so during flight the body can only reposition itself about this point. (If the arms are raised, the body moves downward relative to proportional weights, and if the extremities move forward, the trunk moves backward.) The means for obtaining the total body center of gravity is included in the computer program presented in Appendix B. Determining the body center of gravity also allows one to compare movements of different

The Free Body Diagram for a One-Segment Motion

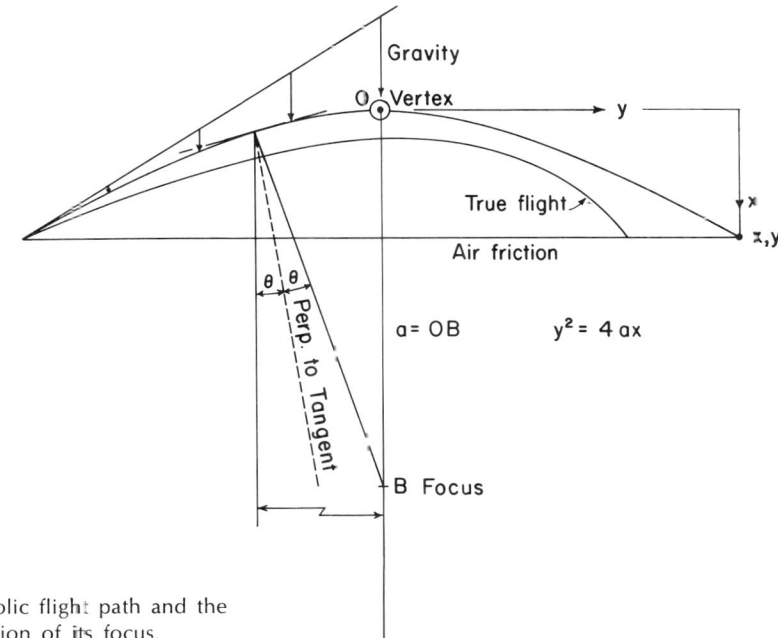

Fig. 4-16
The parabolic flight path and the determination of its focus.

individuals performing the same motion. (What is the path of the center of gravity of a novice and an experienced gymnast performing a kip on the high bar?)

If the joint moments of several body segments are to be analyzed, the free body diagram can still be drawn, but the fixed body segment is now nonexistent. Limb motions can be analyzed relative to the trunk, and the instantaneous velocities and accelerations of the trunk relative to the earth can be included in the force analysis. The analysis of specific, nonfixed-point sports is presented in Chapter 9.

RADIUS OF GYRATION — CENTER OF PERCUSSION

The ladder problem, problem 10 in Appendix A, was used to review the two methods of determining moments of force. When determining joint moments, the perpendicular component of the body segment is multiplied by the distance of the center of gravity from the joint center (r). The moment of force for a one-segment motion was given earlier as:

$$M_o - \text{wt}(\cos\theta)r - (mr\alpha)r = 0$$

However, the moment of a segment is a function of its moment of inertia (I_o), and the moment of inertia is in turn a measure of the resistance of a body to rotation, or if the body is rotating, to coming to rest. It is comparable to mass in a linear problem. It is determined by measuring the time of the period of oscillation of the body when the body is suspended at any desired pivot point. The formula for an irregularly shaped object is used:

The Free Body Diagram for a One-Segment Motion

$$I_o = \frac{t^2 wr}{4\pi^2}$$

where t = period or time of one complete swing (timed for 10 swings and divided by 10);
w = segment weight, and
r = distance from the center of gravity to the pivot point.

Note also that

$$I_o = mk^2$$

where m = mass = w/g and k = radius of gyration.

The radius of gyration is the distance from the axis of rotation to the point about which the mass of a rotating body is concentrated without changing its moment of inertia (I_o). This has proven difficult to understand, so data on a rotating pendulum is presented here to show how the radius of gyration k differs from r, the distance of the center of gravity from the pivot point. A pendulum weighing 925 gm has a period of 1.6 sec, and an r of 55 cm. If I_o must remain unchanged, then the radius of gyration can be determined using the two formulas presented above. If

$$I_o = \frac{t^2 wr}{4\pi^2} \quad \text{and} \quad I_o = mk^2$$

then

$$mk^2 = \frac{t^2 wr}{4\pi^2}$$

Substituting,

$$\frac{925}{980} k^2 = \frac{(1.6)^2(925)(55)}{39.5} \qquad 0.944 k^2 = 3310 \qquad k = 59.2 \text{ cm}$$

Therefore, the distance k must be used (rather than r) for all problems of rotation, and k is always slightly larger than r. Because the force $m r\alpha$ is always perpendicular to the segment, the moment of force due to movement is not $m r\alpha \cdot r$, but $m k\alpha \cdot k$ or $mk^2\alpha$ (or $I_o \alpha$).

Thus the moment of force formula for a one-segment motion in the counterclockwise direction in the first quadrant is:

$$M_o - wt(\cos\theta)r - mk^2\alpha = 0$$

The muscle contraction is pulling the segment in the counterclockwise ($+$) direction and the weight and inertial force is in the clockwise direction ($-$).

A discussion of the radius of gyration is not complete without showing the relationship between distance from the pivot point to the center of gravity (r), radius of gyration (k), and the center of percussion (q) (see Fig. 4-17).

$$q = \frac{k^2}{r}$$

The Free Body Diagram for a One-Segment Motion

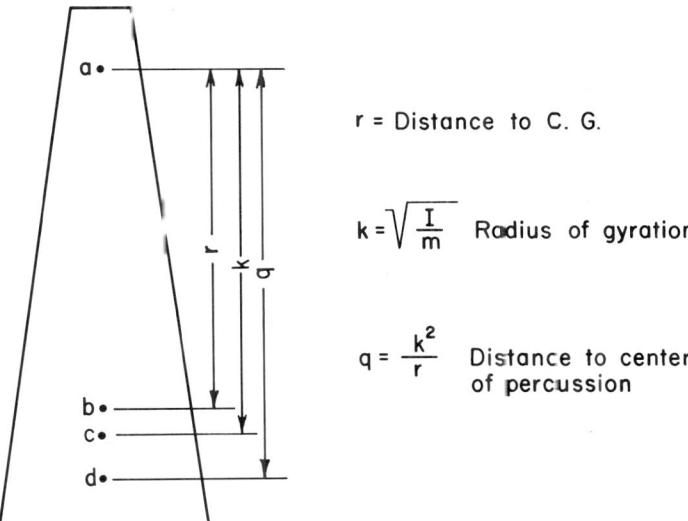

Fig. 4-17
Determination of the radius of gyration and center of percussion.

If an object is swung about point *a* (Fig. 4-17) or about point *d* the period remains the same. Point *d* is called the *center of percussion* since, if the mass is struck at point *d* when pivoting about *a*, the oscillation takes place without producing pressure at *a*. Likewise, if *a* were used as the pivot point, the center of percussion would be at *a*.

The importance of this in sports is readily realized when a ball is struck by a bat at the center of percussion and the effect is felt on the hands. To illustrate the changing forces on the hands as the impact point changes, suspend a bat in mid air using two long wires and strike the bat at various places. If struck at the center of gravity, the whole bat will move equally. If struck at the center of percussion, the bat will rotate with the end of the handle remaining at one point. If struck in either direction away from the center of percussion, a combination of rotation and linear motion occurs. Practical application of these relationships is presented in the discussion of equipment design in Chapter 7.

FORCE AND MOMENT FORMULAS

The force and moment of force formulas can now be written properly for a one-segment motion. They are summarized in Fig. 4-18. The inertial forces $mr\alpha$ and $mr\omega^2$ are shown divided into the X and Y components in all four quadrants with the motion in the counterclockwise direction.

It is customary to make the $\Sigma F = 0$ and the $\Sigma M_o = 0$. This has been changed for the force formulas (F_x and F_y) only as a convenience so the force direction may be shown in the formula in the direction actually seen in the free body diagram. The moment formula is unchanged, because the answer produces the proper direction of the dominant muscles.

The signs of the components of the X and Y forces are obtained by moving from the tail to the head of the vector along the X and Y coordinates. All students should now be able to draw a free body diagram, and be able to write the force and moment formulas for a one-segment motion.

$F_x = -mr\omega^2 \cos\theta + mr\alpha \sin\theta$
$F_y = mr\omega^2 \sin\theta + mr\alpha \cos\theta - wt$
$M_o - mk^2\alpha + wt \cos\theta\, r = 0$

$F_x = mr\omega^2 \cos\theta + mr\alpha \sin\theta$
$F_y = mr\omega^2 \sin\theta - mr\alpha \cos\theta - wt$
$M_o - mk^2\alpha - wt \cos\theta\, r = 0$

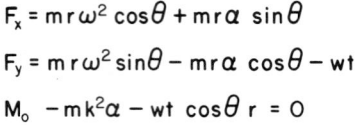
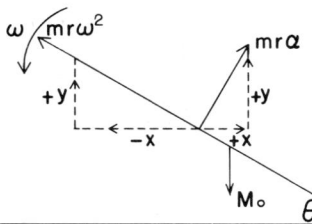
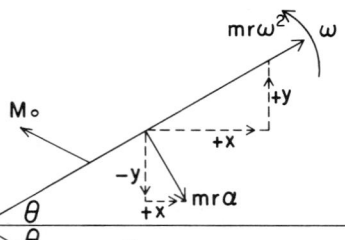
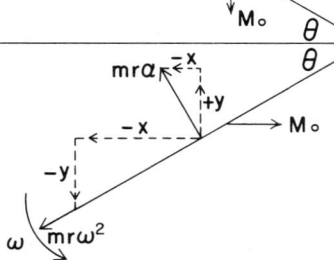
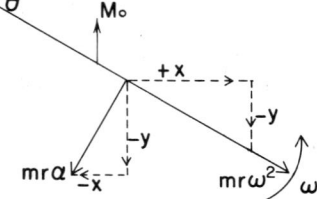

$F_x = -mr\omega^2 \cos\theta - mr\alpha \sin\theta$
$F_y = -mr\omega^2 \sin\theta + mr\alpha \cos\theta - wt$
$M_o - mk^2\alpha + wt \cos\theta\, r = 0$

$F_x = mr\omega^2 \cos\theta - mr\alpha \sin\theta$
$F_y = -mr\omega^2 \sin\theta - mr\alpha \cos\theta - wt$
$M_o - mk^2\alpha - wt \cos\theta\, r = 0$

Fig. 4-18
The forces and moments of forces formulas for a one-segment motion.

The Free Body Diagram for a One-Segment Motion

Glossary of Terms and Symbols
for Drawing a Free Body Diagram

Absolute motion — the total movement of a given particle with respect to a fixed point on the earth. This includes the relative motion of a particle and moving coordinates of that particle.

Alpha (α) — the symbol used for angular acceleration of a body segment.

Angular acceleration (α) — the time rate of change of the angular velocity, or the second time derivative of the displacement of a rotating body segment.

Angular displacement (θ) — the angular position of a line measured from the right horizontal axis.

Angular velocity (ω) — the time rate of change of angular motion of a body segment, or the first time derivative of the displacement.

Center of gravity — the point through which a single upward vector passes that is equal to all downward forces applied to a body by the earth's gravity, putting the body in equilibrium.

Coriolis force — a force produced when a point rotates about a rotating axis. (See Chapter 5).

D'Alembert's principle — all inertial forces in a free body diagram are treated in terms of equal and opposing forces thus reducing a problem in dynamics to a problem of statics. All external forces and inertial forces thus form a system in equilibrium.

External forces — mechanically applied contact forces, gravity, and electrical and magnetic forces which influence body segments.

Free body diagram — the drawing of an isolated body, with force vectors indicated by arrows, showing the state of equilibrium of all forces acting on it.

k — the symbol used for the radius of gyration.

m — the symbol used for mass. It is the weight of a body divided by the acceleration of gravity.

Moment of inertia (I_o) — a measure of the resistance of a body to rotate or, if rotating, to come to rest. The moment of inertia, synonymous with the term torque, is equal to the mass times the radius of gyration squared ($I_o = mk^2$). It is dependent on the distribution of the mass about an axis of rotation, and is found by suspending the body as a physical pendulum and measuring its period of vibration.

Normal force ($mr\omega^2$) — the inertial force due to the rotation of a body segment drawn parallel to the longitudinal axis of the segment from the center of gravity outward from the pivot point.

Omega (ω) — the symbol used for angular velocity of a body segment.

r — the symbol used to designate the distance from the pivot point to the center of gravity of a body segment.

Radius of gyration (k) — the distance from the axis of rotation of a point about which the mass is concentrated in a rotating body without a change in its moment of inertia. ($I_o = mk^2$).

Relative motion — the motion of a body segment with respect to a moving coordinate system (the adjoining segment).

Rigid body — a body segment which may be analyzed as if it were free of deformation.

Rotation — angular motion having all particles moving in a circular path about an axis.

Tangential force ($mr\alpha$) — the inertial force due to the rotation of a body segment, drawn tangent to the longitudinal axis of the segment from the center of gravity

Theta (θ) — the symbol used for the displacement angle of a body segment measured from the right horizontal axis

chapter 5

TWO- AND THREE-SEGMENT MOTIONS

When two segments are in motion, the first is rotating about a fixed point and the second is rotating about the moving end of the first segment (Fig. 5-1). The free body diagram is determined first for the second segment and includes all inertial forces due to the movement of both segments.

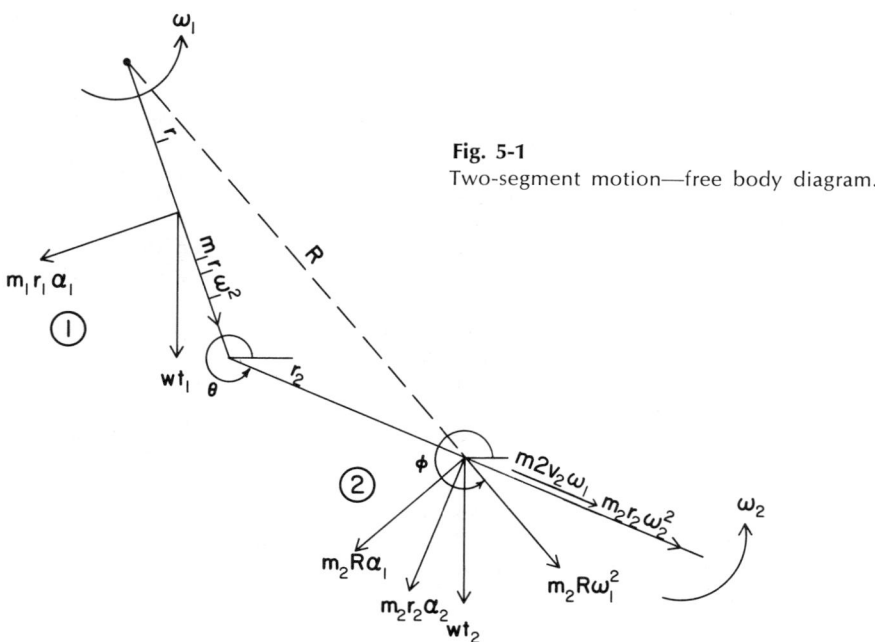

Fig. 5-1
Two-segment motion—free body diagram.

Two and Three-Segment Motions

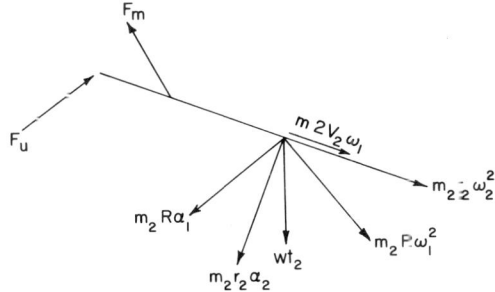

Fig. 5-2
Free body diagram for segment 2.

In the free body diagram of the second segment there are six forces that are in equilibrium with the sum of the forces due to muscular contractions. The resultant force due to total muscle action is indicated by a single vector (F_m) in the free body diagram (Fig. 5-2). The six forces for segment 2 are:

1. Weight of segment 2.
2. The normal force $m_2 r_2 \omega_2^2$. ω is the angular velocity *relative* to segment 1.
3. The tangential force $m_2 r_2 \alpha_2$. α is the angular acceleration *relative* to segment 1.
4. A force due to the movement of segment 1 ($m_2 R \omega_1^2$). With the center of gravity of segment 2 revolving due to the motion of segment 1, a new radius (R) is established, and this force is normal to R.
5. A force due to the movement of segment 1 tangent to the new radius R ($m_2 R \alpha_1$).
6. Coriolis force ($m_2 2 V_2 \omega_1$).

CORIOLIS FORCE

A French scientist, Coriolis, was the first to recognize a force that exists when a segment is rotating about an axis that itself is rotating. This force (for a two-segment motion, $m_2 2 V_2 \omega_1$) is always normal to segment 2 and in the direction in which the head of vector V would move about its tail in the sense of ω_1 (Fig. 5-3). This vector is then drawn in the opposite direction on the free body diagram, as are all inertial forces. The Coriolis force is obtained by equating the absolute and relative accelerations.

Fig. 5-3
Direction of Coriolis acceleration (drawn in opposite direction on free body diagram).

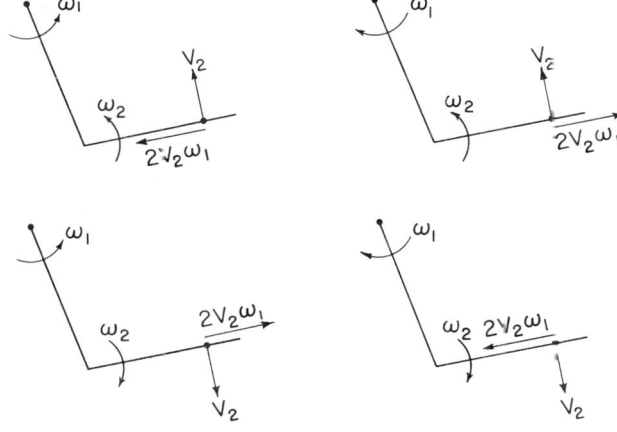

Coriolis Force for the Second Segment of a Two-Segment Motion
(Normal component)

Relative

$$\frac{F}{m} = \omega_2^2 r_2 + \omega_1^2(l_1 + r_2)$$

$$= \omega_2^2 r_2 + \omega_1^2 l_1 + \omega_1^2 r_2$$

Absolute

$$\frac{F}{m} = \omega_1^2 l_1 + (\omega_1 + \omega_2)^2 r_2$$

$$= \omega_1^2 l_1 + (\omega_1^2 + 2\omega_1\omega_2 + \omega_2^2) r_2$$

Absolute acceleration minus relative acceleration = $2\omega_1\omega_2 r_2 = 2\omega_1 V_2$

Coriolis Force for the Third Segment of a Three-Segment Motion
(Normal component)

Relative

$$\frac{F}{m} = \omega_3^2 r_3 + \omega_2^2(l_2 + r_3) + \omega_1^2(l_1 + l_2 + r_3)$$

$$= \omega_3^2 r_3 + \omega_2^2 l_2 + \omega_2^2 r_3 + \omega_1^2 l_1 + \omega_1^2 l_2 + \omega_1^2 r_3$$

Absolute

$$\frac{F}{m} = \omega_1^2 l_1 + (\omega_1 + \omega_2)^2 l_2 + (\omega_1 + \omega_2 + \omega_3)^2 r_3$$

$$= \omega_1^2 l_1 + l_2\omega_1^2 + l_2 2\omega_1\omega_2 + l_2\omega_2^2 + r_3\omega_1^2 + r_3\omega_2^2 + r_3\omega_3^2 + r_3 2\omega_1\omega_3 + r_3 2\omega_2\omega_3$$
$$+ r_3 2\omega_1\omega_2$$

$$= l_2 2\omega_1\omega_2 + r_3 2\omega_1\omega_3 + r_3 2\omega_2\omega_3 + r_3 2\omega_1\omega_2$$

Absolute minus relative acceleration = $2\omega_1 V_2 + 2\omega_1 V_3 + 2\omega_2 V_3 + 2\omega_1\omega_2 r_3$

Coriolis Force for the Fourth Segment of a Four-Segment Motion
(Normal component)

$$\frac{F}{m} = 2\omega_1 V_2 + 2\omega_1 V_3 + 2\omega_2 V_3 + 2\omega_1\omega_2 l_3 + 2\omega_1\omega_2 r_4 + 2\omega_1\omega_3 r_4 + 2\omega_1 V_4$$
$$+ 2\omega_2\omega_3 r_4 + 2\omega_2 V_4 + 2\omega_3 V_4$$

The free body diagram for segment 2 can now be completed showing the six forces, the resultant muscle force (F_m) which produced the motion, and the push of the first segment against the end of the second segment (F_u) (Fig. 5-2). The actual direction and magnitude of the bone-on-bone joint force (F_u) cannot be determined, so the vector indicates only the existing force.

The free body diagram of segment 1, of the two-segment motion, is shown in Fig. 5-4. The five forces are:

1. Weight of segment 1.
2. The normal force due to the motion of segment 1 about the fixed axis ($m_1 r_1 \omega_1^2$).
3. The tangential force due to the motion of segment 1 ($m_1 r_1 \alpha_1$).
4. The sum of the forces of segment 2 applied to the distal end of segment 1.

5. The moments of force of segment 2 applied in the opposite direction to segment 1. (The resultant muscle force of segment 2 also attempts to rotate segment 1 in the opposite direction.)

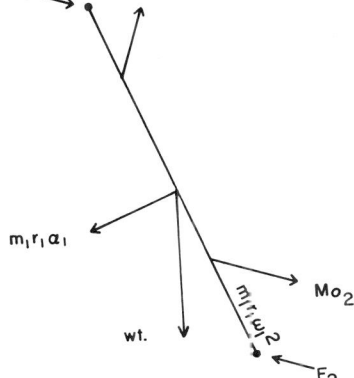

Fig. 5-4
Free body diagram for segment 1.

The Force and Moment Formulas for the Two-Segment Motion of Fig. 5-1
(Also, see Appendix C)

Segment 2

$$F_{y_2} = -WT_2 - m_2r_2\alpha_2 \cos(360° - \theta_2) - m_2r_2\omega_2^2 \sin(360° - \theta_2) - m_2R\omega_1^2 \sin(360° - \phi) - m_2R\alpha_1 \cos(360° - \phi) - m2V_2\omega_1 \sin(360° - \theta_2)$$

$$F_{x_2} = -m_2r_2\alpha_2 \sin(360° - \theta_2) + m_2r_2\omega_2^2 \cos(360° - \theta_2) + m_2R\omega_1^2 \cos(360° - \phi) - m_2R\alpha_1 \sin(360° - \phi) + m2V_2\omega_1 \cos(360° - \theta_2)$$

$$M_{o_2} - WT_2 \cos(360° - \theta_2)r_2 - m_2k_2^2\alpha_2 - m_2R\omega_1^2 \sin(\theta_2 - \phi)r_2 - m_2R\alpha_1 \cos(\theta - \phi)r_2 = 0$$

Segment 1

$$F_{y_1} = -WT_1 - m_1r_1\omega_1^2 \sin(360° - \theta_1) - m_1r_1\alpha_1 \cos(360° - \theta_1) + F_{y_2}$$

$$F_{x_1} = m_1r_1\omega_1^2 \cos(360° - \theta_1) - m_1r_1\alpha_1 \sin(360° - \theta_1) + F_{x_2}$$

$$M_{o_1} - WT_1 \cos(360° - \theta_1)r_1 - m_1k_1^2\alpha_1 \pm F_{y_2}l_1 \cos(360° - \theta) \pm F_{x_2}l_1 \sin(360° - \theta) - M_{o_2} = 0$$

The relative motion method of analysis is presented, rather than the absolute method, so the forces due to the five accelerations are clearly separated. [A governor mechanism problem using the relative motion method is given by Ham and Crane (C-12, p. 298).] The absolute method allows for easier calculation, but the forces due to the separate motion of all segments are not defined. The relative motion method shows how every body segment influences every other body segment. The ability to interpret the forces and moments obtained

Two and Three-Segment Motions

allows a more critical analysis of motion. Timing of the relative acceleration of each body part and the magnitude of the moments of force indicate the motion's efficiency. The contribution of each body segment to the whole motion may also be found. This is obtained in the computer program, Appendix B, when the relative velocity and acceleration for a given segment is zeroed, and the whole motion recalculated to find out how much change occurs without the movement of that particular segment. The velocities and accelerations are reinserted and the entire analysis is repeated with the next segment velocities and accelerations zeroed. In this manner the relative contribution of each segment's movement can be determined. This could not be done using the absolute motion method. (The absolute motion method is presented in Appendix C.)

A three-segment motion analysis and the use of the computer programs should be reserved for graduate students. The undergraduate should be aware of the numerous forces due to motion and the complexity of the calculations without being responsible for determining force magnitudes and directions. When the study of three-segment motion is completed, students realize fully that muscle action is totally unpredictable from observation of movements alone.

THREE-SEGMENT MOTION

Figure 5-5 shows a three-segment motion with segment 1 rotating about a fixed point, and segments 2 and 3 rotating about a moving axis. (Note segments 2 and 3 have a minus angular acceleration.) The free body diagram for each segment, showing inertial forces and weight, is presented in Fig. 5-6, and Fig. 5-7 gives a breakdown of the forces to aid in writing the force formulas. The force and moment formulas are as follows:

Segment 3

$$F_{y_3} = -WT_3 + m_3 r_3 \alpha_3 \cos \theta_3 + m_3 r_3 \omega_3^2 \sin \theta_3 - m_3 R_1 \alpha_1 \cos \phi_1 + m_3 R_1 \omega_1^2 \sin \phi_1$$
$$+ m_3 R_2 \alpha_2 \cos (180° - \phi_2) + m_3 R_2 \omega_2^2 \sin (180° - \phi_2)$$
$$+ m_3 (2\omega_1 V_2 + 2\omega_1 V_3 + 2\omega_1 \omega_2 r_3) \sin \theta_3 - m2 \omega_2 V_3 \sin \theta_3$$

$$F_{x_3} = -m_3 r_3 \alpha_3 \sin \theta_3 + m_3 r_3 \omega_3^2 \cos \theta_3 + m_3 R_1 \alpha_1 \sin \phi_1 + m_3 R_1 \omega_1^2 \cos \phi_1$$
$$+ m_3 R_2 \alpha_2 \sin (180° - \phi_2) - m_3 R_2 \omega_2^2 \cos (180° - \phi_2)$$
$$+ m_3 (2\omega_1 V_2 + 2\omega_1 V_3 + 2\omega_1 \omega_2 r_3) \cos \theta_3 - m2 \omega_2 V_3 \cos \theta_3$$

$$M_{o_3} - WT_3 \cos \theta_3 r_3 + m_3 k_3^2 \alpha_3 + m_3 R_1 \omega_1^2 \sin (\phi_1 - \theta_3) r_3 - m_3 R_1 \alpha_1 \cos (\phi_1 - \theta_3) r_3$$
$$+ m_3 R_2 \omega_2^2 \sin (\phi_2 - \theta_3) r_3 - m_3 R_2 \alpha_2 \cos (\phi_2 - \theta_3) r_3 = 0$$

Segment 2

$$F_{y_2} = -WT_2 + m_2 r_2 \alpha_2 \cos (180° - \theta_2) + m_2 r_2 \omega_2^2 \sin (180° - \theta_2)$$
$$- m_2 R_1 \alpha_1 \cos \phi_1^1 + m_2 R_1 \omega_1^2 \sin \phi_1^1 - m2 V_2 \omega_1 \sin (180° - \theta_2) + F_{y_3}$$

$$F_{x_2} = +m_2 r_2 \alpha_2 \sin (180° - \theta_2) - m_2 r_2 \omega_2^2 \cos (180° - \theta_2) + m_2 R_1 \alpha_1 \sin \phi_1^1$$
$$+ m_2 R_1 \omega_1^2 \cos \phi_1^1 + m2 V_2 \omega_1 \cos (180° - \theta_2) + F_{x_3}$$

$$M_{o_2} + WT_2 \cos (180° - \theta_2) r_2 - m_2 k_2^2 \alpha_2 - m_2 R_1^1 \omega_1^2 \sin (\theta_2 - \phi_1^1) r_2$$
$$- m_2 R_1^1 \alpha_1 \cos (\theta_2 - \phi_1^1) r_2 + F_{y_3} l_2 (\cos 180° - \theta_2)$$
$$+ F_{x_3} l_2 (\sin 180° - \theta_2) - M_{o_3} = 0$$

Two and Three-Segment Motions

Segment 1

$$F_{y_1} = -WT_1 - m_1 r_1 \alpha_1 \cos\theta_1 + m_1 r_1 \omega_1^2 \sin\theta_1 + F_{y_2}$$

$$F_{x_1} = +r_1 r_1 \alpha_1 \sin\theta_1 + m_1 r_1 \omega_1^2 \cos\theta_1 + F_{x_2}$$

$$M_{o_1} - WT_1 \cos\theta_1 r_1 - m_1 k_1^2 \alpha_1 + F_{y_2} l_1 (\cos\theta_1) + F_{x_2} l_1 (\sin\theta_1) - M_{o_2} = 0$$

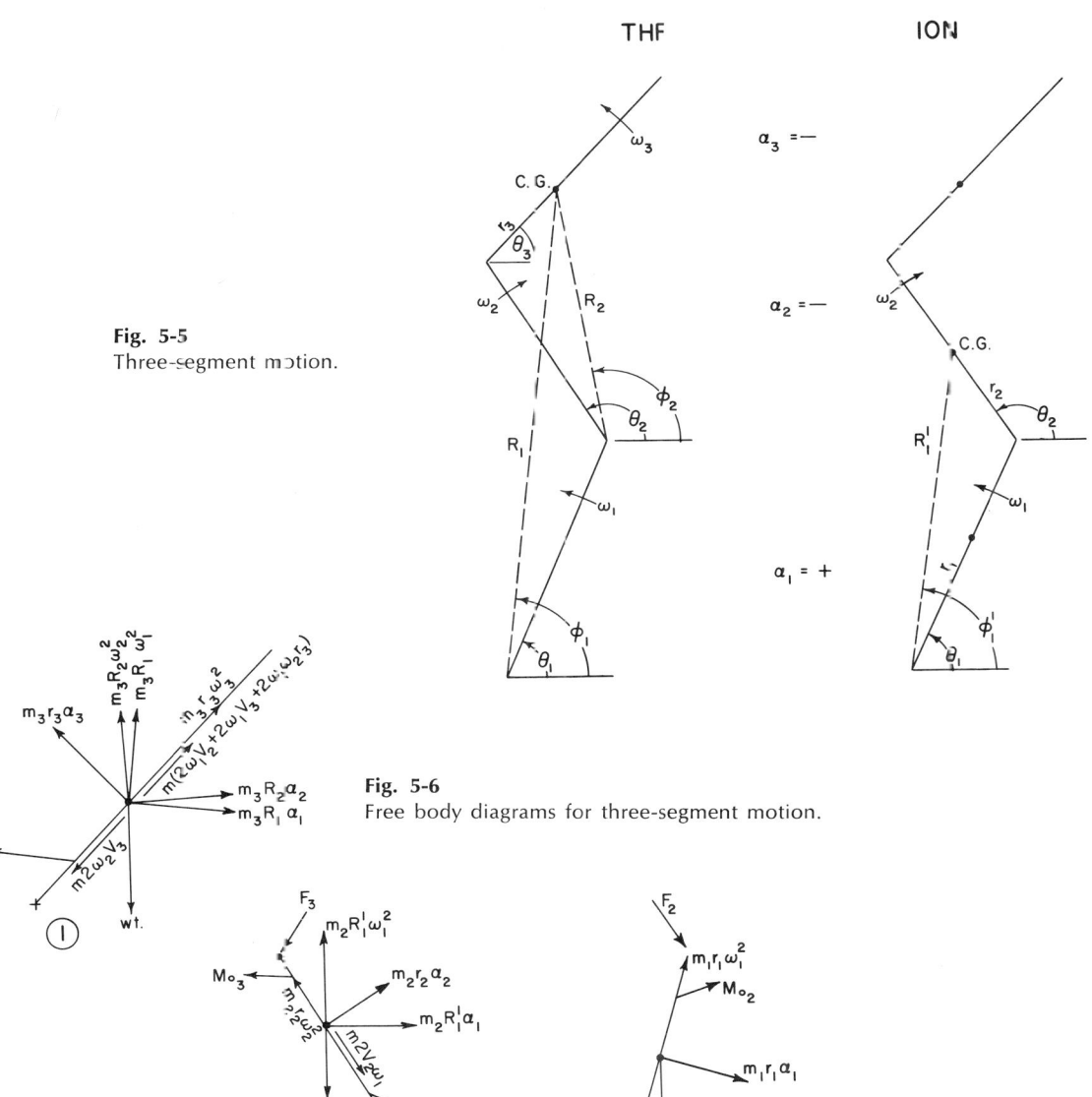

Fig. 5-5
Three-segment motion.

Fig. 5-6
Free body diagrams for three-segment motion.

Two and Three-Segment Motions

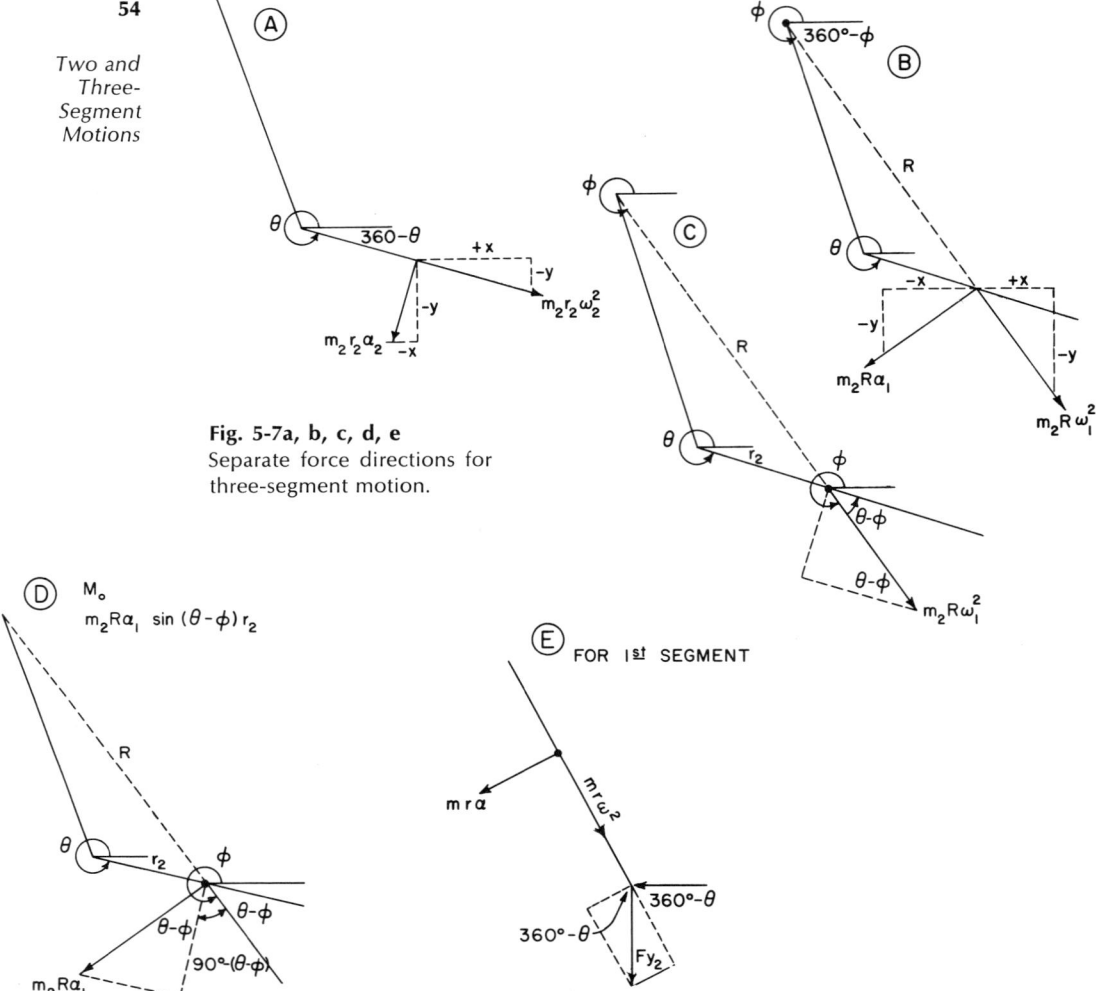

Fig. 5-7a, b, c, d, e
Separate force directions for three-segment motion.

The three-segment motion has eight inertial forces for the third segment, five for the second segment, and two for the first segment. If a movement for five segments were analyzed (Fig. 5-8), there would be 38 inertial forces due to motion of all the body parts. It can readily be seen at this point that the movement of every body segment affects every other segment, and involves muscle action at every body joint. It is also obvious that even a three-segment motion would take such a long time to analyze by hand that it would be impractical. The computer has made such an analysis practical. A program that obtains the segment velocities and accelerations, the horizontal and vertical forces, the moments of force, and the contribution of each segment toward the total motion is presented in Appendix B. Using this information, not only can a specific motion be analyzed, but the quality of the performance may be ascertained. This can be done by listing the positions of maximum absolute deceleration of each body segment, the extent of the muscle action at each

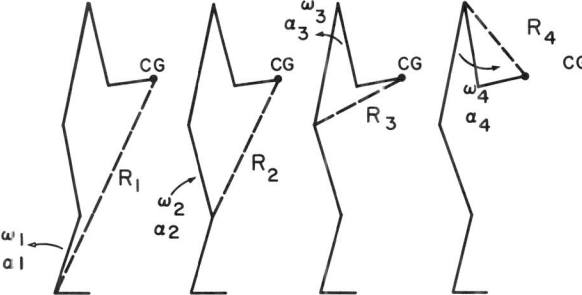

Fig. 5-8
Five-segment motion—movement of forearm due to movement of other body segments.

joint, and the extent of contribution of each body segment. The movement of the segment nearest the fixed point should accelerate and increase velocity; then the deceleration of that first segment aids the increase in velocity of the next segment. The same sequence takes place so that the deceleration of segment 2 aids the increase in velocity of segment 3. The timed sequence of one segment helping the next will produce the maximum velocity of the last segment with a minimum of muscle force at each joint. By noting the timing of velocities, accelerations, and decelerations of each segment and the magnitude of the joint moments, the efficiency of motion may be interpreted. (Compare a kip performed on a high bar by a novice and by an advanced gymnast. The gymnast's proper timing will result in less muscle force at each joint.)

INTERPRETING MOMENTS

The moments indicate not only dominant muscle forces; the moment formula also indicates the effect of one segment on the adjoining segment. If muscle action is moving a segment in the clockwise direction, it will be attempting to move the adjoining segment in the counterclockwise direction; thus the moment formula has the moments from the preceding segment applied in the opposite direction. At times, the moments of one segment are so large that they will be the dominating muscle force at the next segment, producing a dominant muscle action at this adjoining segment just opposite to that which might be expected. This is true in the standing broad jump where hip extension is so great that the knee flexors are dominant even though the legs are straightening. To illustrate this point, Fig. 5-9a–d shows four static body positions for comparison.

Deep Knee Bends with a 100 lb Barbell

Figure 5-9a presents a deep knee bend with the thighs parallel to the floor, but with the trunk bent forward 35° from the horizontal. The moment analysis shows that the dominant muscle actions are: hip extension, $-1,843,792$ gm-cm; knee extension, 699,038 gm-cm; and ankle extension, $-257,153$ gm-cm.

Figure 5-9b illustrates a position with the trunk almost upright, but with the thigh remaining close to the horizontal. The dominant muscle actions are: hip extension, $-765,432$ gm-cm; knee extension, 1,751,969 gm-cm; and ankle extension, $-47,921$ gm-cm. A comparison of Fig. 5-9a and b shows that an increase in the magnitude of knee extension occurs when the magnitude of hip

Two and Three-Segment Motions

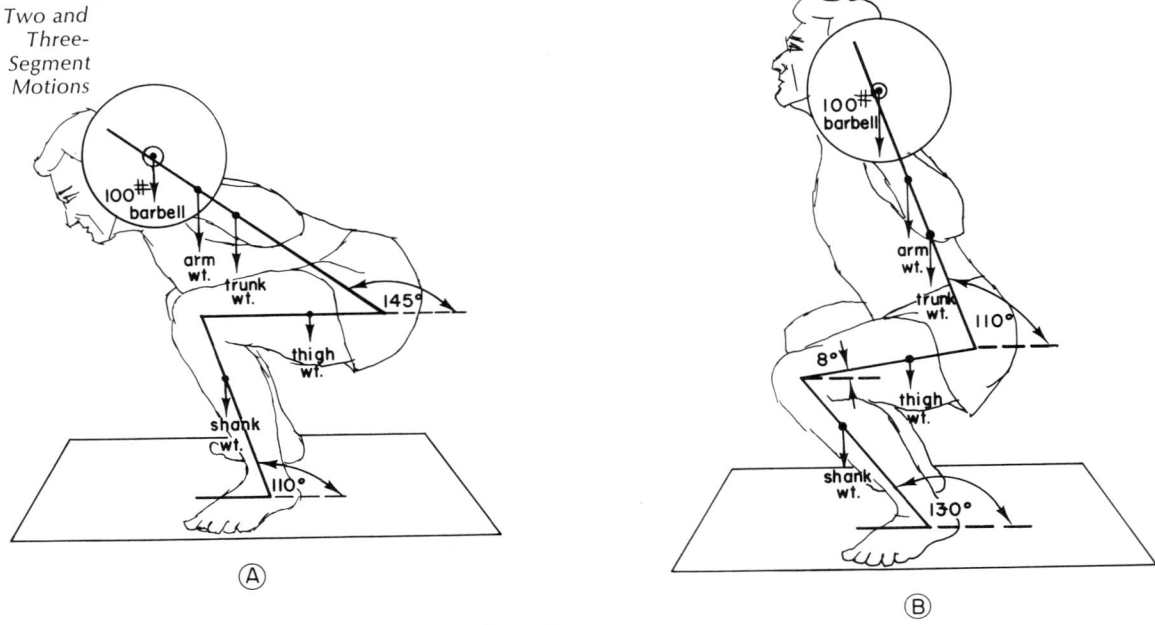

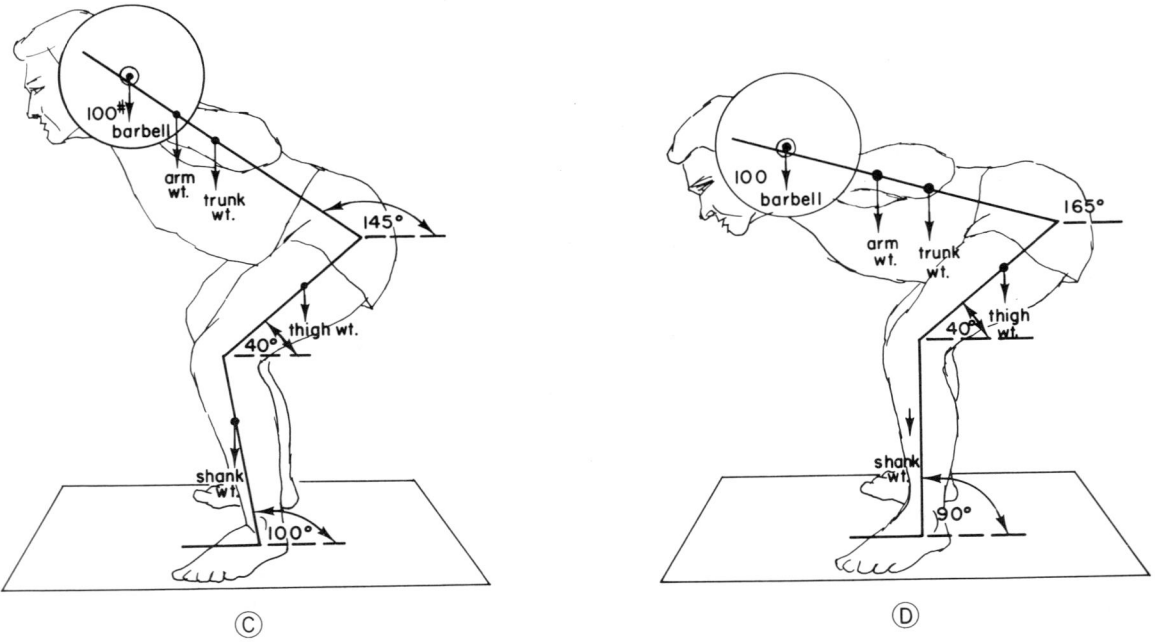

Fig. 5-9a, b, c, d
Static deep knee bend positions for analysis to show the changing muscle action (joint moments of force) with changing positions.

Two and Three-Segment Motions

extension decreases, and that the muscle action at the ankle remains relatively small in both positions.

Figure 5-9c has the trunk position as in Fig. 5-9a, but the legs are partially straightened so the thigh is 40° from the horizontal. The dominant muscle actions are: hip extension, $-1,843,792$ gm-cm; knee extension, $104,015$ gm-cm; and ankle extension, $-385,330$ gm-cm. This position is used to illustrate the fact that large moments at the hip resulting in relatively small moments at the knee, due to thigh position, result in large moments at the ankle joint.

Figure 5-9d has the thigh position as in Fig. 5-9c, but the trunk is bent forward even more (15° from the horizontal). The dominant muscle actions are: hip extension, $-2,174,728$ gm-cm; knee flexion, $-226,921$ gm-cm; and ankle extension, $-226,921$ gm-cm. The large moments at the hip produce moments at the knee opposite to those of Fig. 5-9a, b, and c, and opposite to what is generally expected when simply viewing the position. If the knee flexors were not being used, the static position would not be maintained, and the legs would be straightening. When the forces due to motion are a part of an analysis, the predicted dominant muscle action is almost indeterminant. These static positions are used to clarify the fact that the dominant muscle action at a joint may be opposite to the motion of the joint (knee flexors dominant while knee extension is taking place).

These four static positions should be hand calculated (slide rule) to fully understand how the moments at one joint affect the moments at all the other joints. The necessary data for the analyses are as follows:

Table 5-1 Data for Static Deep Knee Bend Problems

Segment	r (cm)	l (cm)	Weight (gm)	
3	45.6		49,370	½ trunk and barbell and one arm
2	26.8	47	8,300	one thigh
1	26.8	47	3,800	one shank

If extensions of the ankles, knees, and hips are coordinated in a precise pattern during a motion, the moments at each joint can be kept at a low, safe level. However, if knee extension precedes back extension slightly, the use of the hip extensors and ankle extensors would be greater. This type of analysis, therefore, could shed light on the reasons for muscle injuries such as strains, hamstring "pulls," or the Achilles tendon rupture. All motions that have a fast start or stop should be analyzed to show either how injury occurs, or how injury may be avoided by using the proper combination of body segment motions to reduce the joint moments (track starts; running fast; starts and stops in tennis, squash, badminton, handball, baseball, football, etc.).

chapter 6

EXTERNAL FORCES

We have presented the techniques and methods used to analyze human motion, and shall now undertake the analysis of a wide variety of specific motions. The one- or two-segment symmetrical motions should not present a problem as long as no external forces are introduced. Therefore, our emphasis will switch to obtaining tracing practice on motions that are nonsymmetrical and nonplanar.

Subjective error in determining the joint centers can be overcome only by tracing several difficult motions, e.g., the golf drive, soccer kick, or tennis serve. The golf drive must have the right and left sides analyzed separately; the soccer kick analysis moves from the stationary nonkicking foot up across the hips and down the kicking leg, with the movement of the upper body and arms disregarded. The tennis serve measurements progress from the left foot up across the hips and up the right side to the racket hand, with the left arm and right leg disregarded. Motions in the air (soccer head, Mawdsley I-56) or in the water must be analyzed using a moving joint (hip) as the starting point.

The problem of including the whole body in an analysis of a motion such as the tennis serve has been solved by considering the forces due to the motion of the disregarded body segments as external forces to be applied to the link system analyzed. The left arm and right leg of the tennis serve are analyzed separately, and the forces applied at the appropriate joint. The computer program for this procedure is presented in Appendix B. This program also makes it possible to apply the external forces due to impact on the appropriate body segment, or to apply the constant external forces on the body as they occur in rowing or bicycling.

FORCE OF IMPACT

External Forces

In all sports where collisions occur the force of impact must be measured. Impact in sports falls into three main categories:

1. Impact directly on the body by an object of known mass (soccerball, volleyball, handball)
2. Impact of a held implement with an object of known mass (tennis, golf, squash, badminton), and
3. Impact directly on the body by an unknown mass (football, hockey, lacrosse).

The force of impact is obtained by dividing the time of the impact into its momentum ($F = mv/t$). One technique for measuring the time of impact for various balls is the high-speed motion picture camera. Table 6-1 presents data obtained from motion pictures taken at 4000 f/s. Additional information on ball velocities is presented for problems involving force of impact and conservation of momentum.

Table 6-1 Ball and Striker Velocities Before and After Impact; Times of Impact

	BALL VELOCITY		STRIKER VELOCITY		TIME OF IMPACT
	Before	After (ft/sec)	Before (ft/sec)	After (ft/sec)	(sec)
Baseball	hit from tee	128	103	90	$\frac{1}{800}$
Badminton	30 ft/sec	112	97.2	95	$\frac{1}{800}$
Football, punt	0	92	60	39	$\frac{1}{125}$
kickoff	0	85	60	47	$\frac{1}{125}$
Golf, 7 iron	0	177	155	133	$\frac{1}{800}$
drive	0	225	166	114	$\frac{1}{1000}$
Handball serve	0	76	63	47	$\frac{1}{80}$
Paddleball	0	115	90	70	$\frac{1}{200}$
Soccer, kick	0	85	58	42	$\frac{1}{125}$
head	25	42	11	8.6	$\frac{1}{44}$
Squash, hard serve	0	160	145	111	$\frac{1}{333}$
Softball	hit from tee	100	105	71	$\frac{1}{285}$
Tennis, serve	0	167	123	107	$\frac{1}{250}$
forehand	0	93	71	41	$\frac{1}{200}$
Volleyball serve	0	72	65	45	$\frac{1}{100}$

When the external force due to impact is determined, it can be applied directly to the analysis for that position when impact takes place. However, only impacts directly on the body can be used in this manner. When an object or ball is held in the hand and released, it must be considered as part of the last segment, and the center of gravity and moment of inertia for the segment must be calculated. If the object is considered an external force, the inertial forces before release would be eliminated (shot, javelin, discus, baseball, basketball, football).

External Forces

For sports such as tennis, golf, squash, badminton, and baseball batting, the force of impact on the hand(s) is reduced due to partial absorption of force by the implement. The force that reaches the hand varies with design and materials used. Measurements of this transfer of force of impact can be obtained using a flat force transducer placed between the hand and the implement. The strain gauge technique has been used in rowing by Ishiko (I-44), and his results are presented in Fig. 6-1a and b. Similar measurements must be obtained in many other sports to make a complete motion analysis possible.

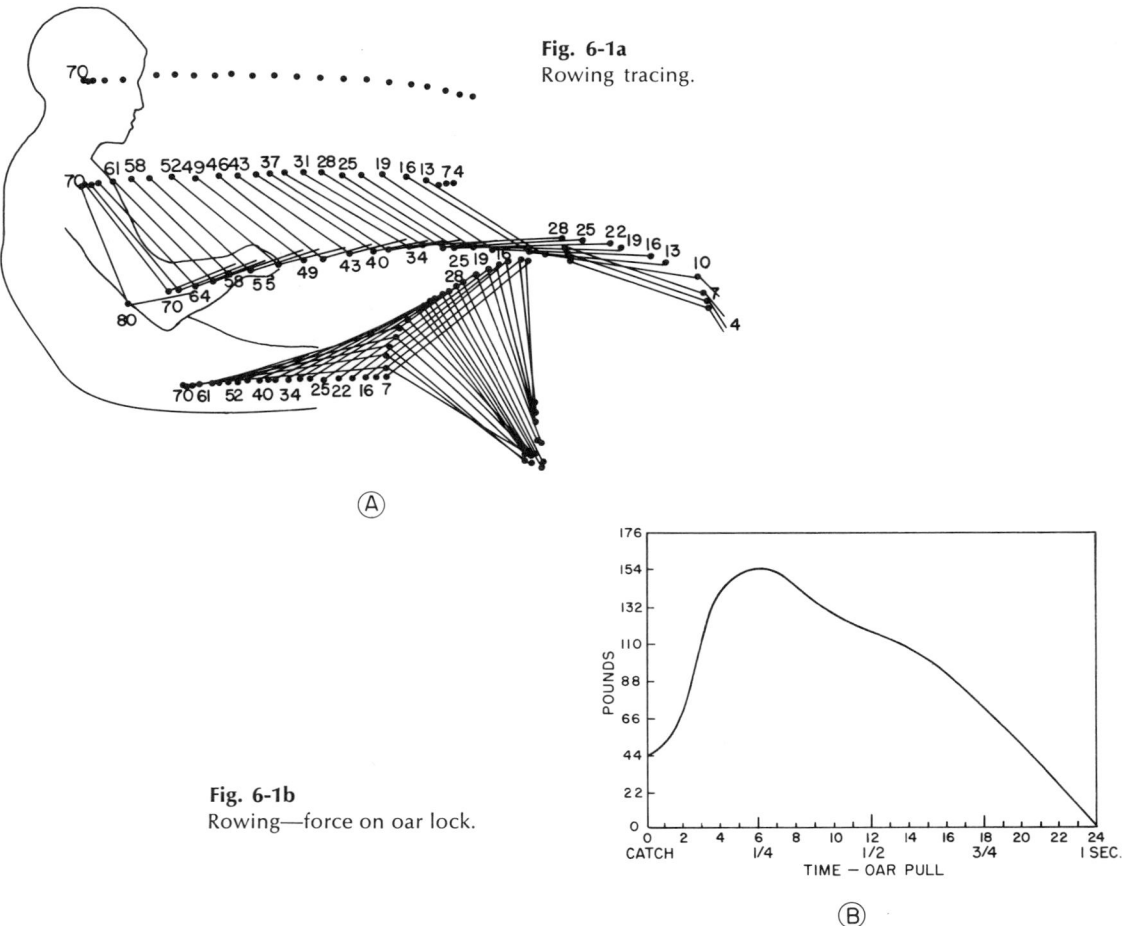

Fig. 6-1a
Rowing tracing.

Fig. 6-1b
Rowing—force on oar lock.

The measurement of the force of impact is especially difficult in the body contact sports. To analyze a shoulder block (Fig. 6-2), strain gauge equipment is required to obtain the external force applied at the shoulder. As determined by Elbel, Wilson, and French (I-22), among 45 players the average force of impact was 267 lb. By applying this figure as an external force, the joint by joint moment analysis can be completed.

Fig. 6-2
Football shoulder block.

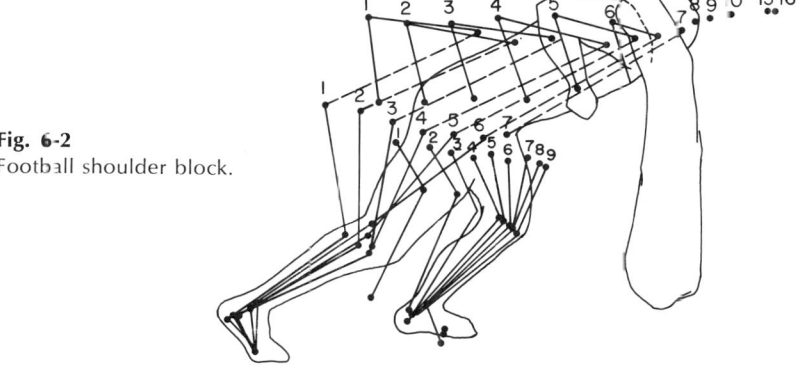

THE STRIKING MASS DURING IMPACT

An important variable in understanding sports proficiency is the striking mass, which is calculated from the formula for conservation of momentum:

$$m_1V_1 + m_2V_2 = m_1V_3 + m_2V_4$$

Because the velocities are measured from motion pictures the coefficient of restitution is not required. We can solve for the striking mass of a golf drive, given ball and club velocity before and after impact and ball weight, as follows:

Before impact:
  Ball $V_1 = 0$
  Club $V_2 = 166$ ft/sec
After impact:
  Ball $V_3 = 225$ ft/sec
  Club $V_4 = 114$ ft/sec
  Ball weight = 47 gm

$$m_1V_1 + m_2V_2 = m_1V_3 + m_2V_4$$
$$(47 \times 0) + (m_2 \times 166) = (47 \times 225) + (m_2 \times 114)$$
$$m_2 = 203.4 \text{ gm}$$

This procedure was followed on measurements taken of golf drives by both men and women professionals on the practice tee at Sutton, Massachusetts, in 1969. The subjects were Floyd, Jacklin, Weiskopf, Marr, Sanders, Palmer, Henning, Beard, Still, and Stockton for men; and Mann, Wright, Whitworth, Haynie, Whalen, and Rawls for women. The results showed that the fastest ball velocity after impact was 185 mph, with the average for men at 160 mph. The fastest ball velocity for women was 158 mph, with an average of 125 mph. The men also moved the club head 24 mph faster than the women before impact, and the greater grip strength of the men resulted in an 84 gm advantage in the striking mass. The ranges indicated that striking mass is a more variable parameter than is club head velocity, thus emphasizing the importance of grip position and firmness.

The same measurements were also obtained with male professional tennis players (I-66), and the results also indicated great variations in the striking mass. As the racket head velocity increases it becomes more difficult to hold

62

External Forces

the grip firmly. As a result, lower club velocities usually allow a greater striking mass than do higher club velocities. This indicates the importance of a balance between club head velocity and striking mass in obtaining maximum ball velocity.

Striking mass changes with the part of the body used and with the individual's ability to maintain body rigidity at impact. The striking mass in handball is approximately 380 gm when the ball is hit coming off the front wall, but about 900 gm when hit off the back wall. During a kicking motion in soccer or football, the striking mass is approximately 2500 gm, again depending on foot velocity and rigidity.

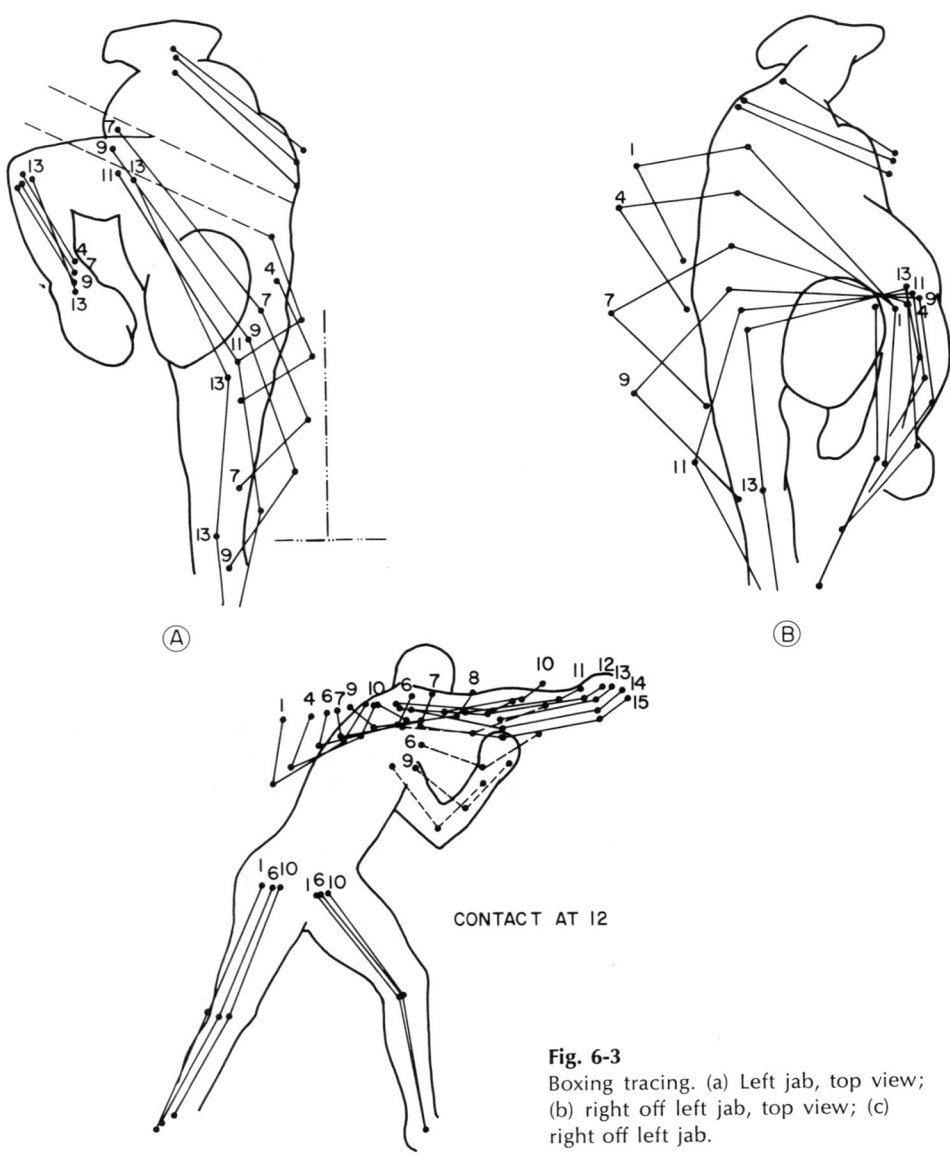

Fig. 6-3
Boxing tracing. (a) Left jab, top view; (b) right off left jab, top view; (c) right off left jab.

Striking Masses in Karate and Boxing

External Forces

The greatest striking mass the body can deliver is produced in the sports of karate and boxing. The striking mass of a boxer's straight right (hitting a 40 lb bag) was measured at 4500 gm (see Fig. 6-3a–c). This motion is similar to the horizontal striking blow using the knuckles in karate (Fig. 6-4a–c) The striking mass could not be determined using the conservation of momentum technique, but pieces of wood were broken using the vertical punch (Fig. 6-5a and b), the edge of the hand chop (Fig. 6-6a and b), and the foot break (Fig. 6-7a and b). The strength required to break the wood was very close to the

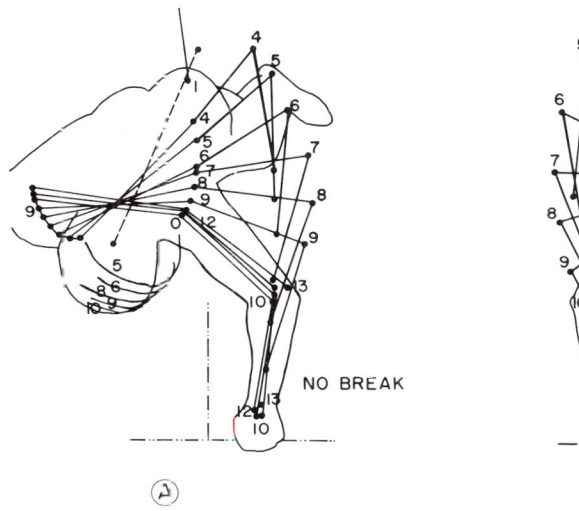
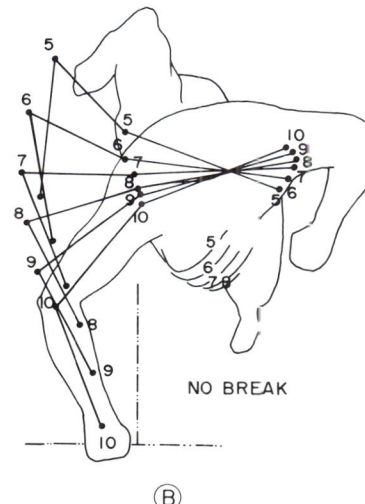

Fig. 6-4
Karate tracing—horizontal punch.
(a) and (b) top view; (c) horizontal view.

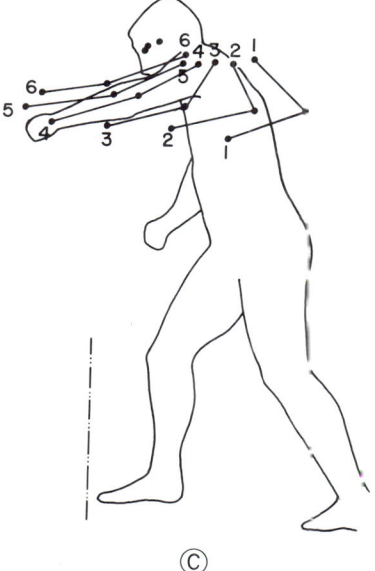

External Forces

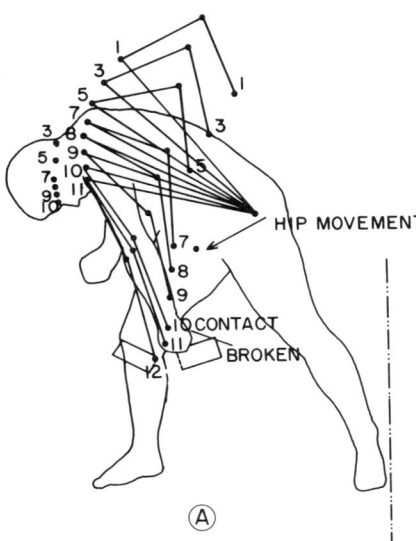
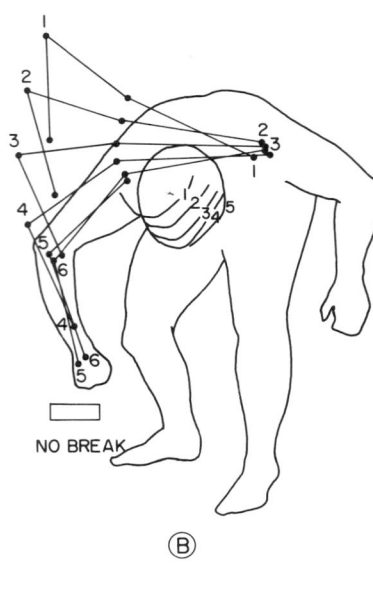

Fig. 6-5 a, b
Karate tracing—vertical punch.

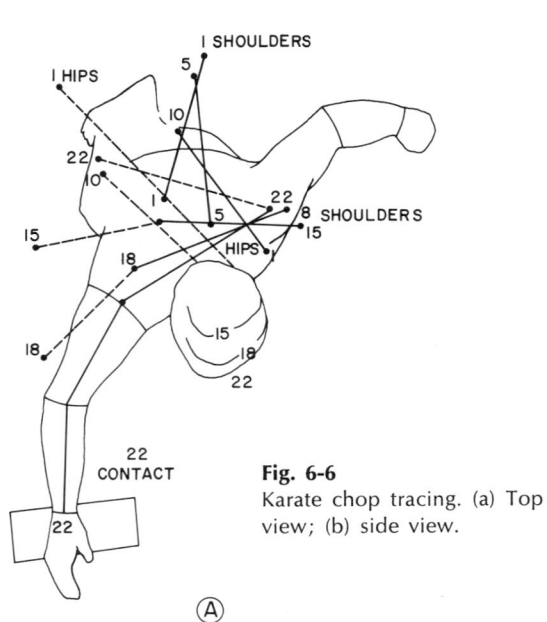

Fig. 6-6
Karate chop tracing. (a) Top view; (b) side view.

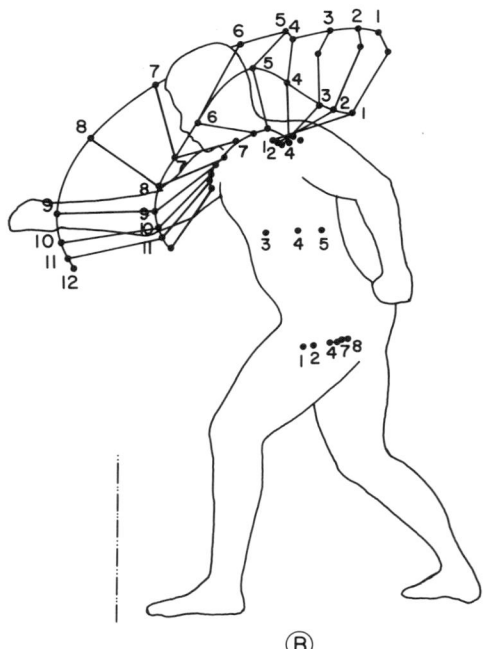

limit of the individual's strength in the punch, while the wood was broken with greater ease using the chop and foot breaks. The wood used was a 2" x 4" dry kiln pine cut into several pieces 12" long. A barbell with only the surface area of the bar lying across the wood was used to break one of the pieces. This static loading required 500 lb to cause rupture. Since the ratio of gradual loading to the sudden loading at impact is approximately 1 to $1\frac{3}{4}$, the approximate total force needed for rupture was 875 lb (174,800 gm). The wood was not broken when the line of hand movement was not perpendicular to the surface of the wood (see Figs. 6-4 and 6-5).

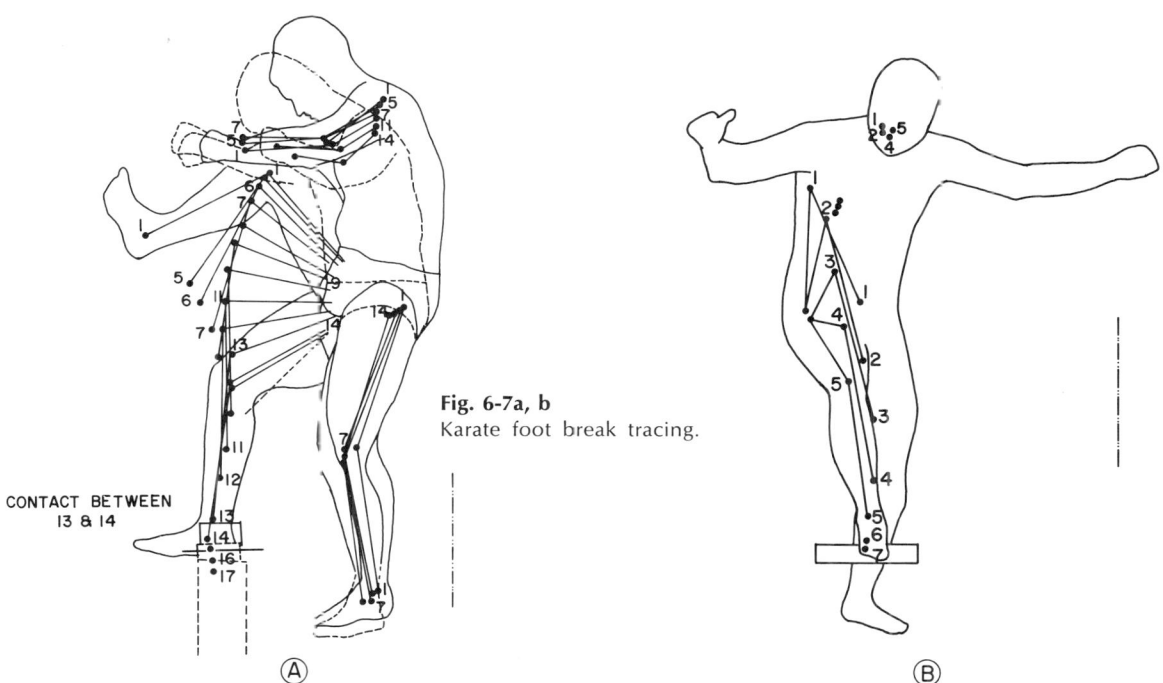

Fig. 6-7a, b
Karate foot break tracing.

It is interesting to note the degree of trunk rotation and to compare the pivot points of the trunk in the top views of boxing and karate. In addition to the alignment of the striking force, the trunk rotates around a point near the opposite shoulder, but in the no-break punch of Fig. 6-4b the rotation point is closer to the mid-line of the trunk. Since whole body force and moment analysis depends on the accurate measurement of external forces when they exist, it is evident that methods for such measurement are necessary.

Body Rigidity

The ability to attain body rigidity during impact is the key to better performance in many sports. Power hitting in handball and the hard kick of a soccer ball depend on firmness of contact. Figure 6-8a–c illustrates the well hit handball, while Fig. 6-9a and b shows a ball struck high on the edge of the

External Forces

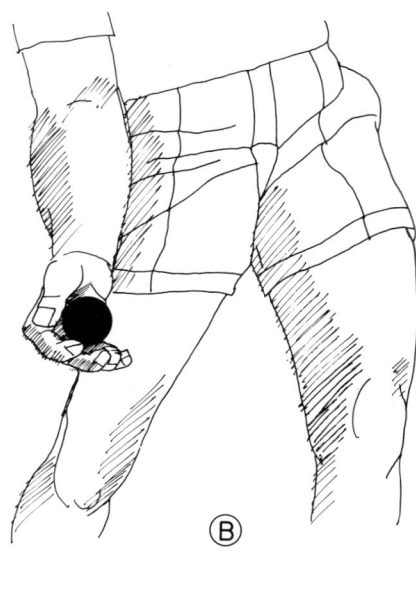

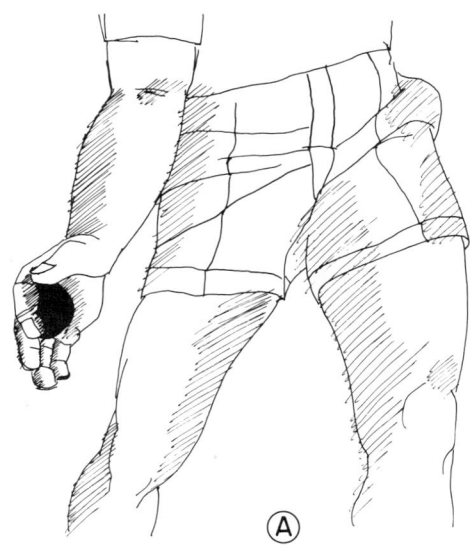

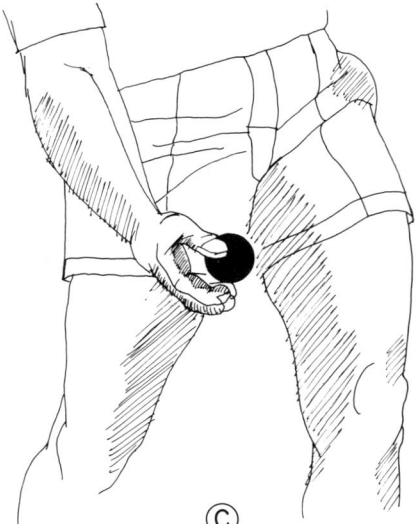

Fig. 6-8a, b, c
The well-hit handball from motion pictures taken at 2000 f/s.

index finger. The latter ball is struck only about one inch away from the desired spot at the base of the index finger, but the result is a greatly reduced ball velocity. The index finger is driven backward due to the impact resulting in a decreased striking mass. In addition, the hand moves ahead of the ball as the ball slides over the edge of the finger. Figure 6-10 shows the solidly hit soccer ball with the foot fully extended at impact, so as to minimize backward movement. Figure 6-11 shows a poorly placed foot striking the ball off center,

Fig. 6-9a, b
A poorly contacted handball from motion pictures taken at 2000 f/s.

Fig. 6-10
Solidly contacted soccer kick.

Fig. 6-11
Poorly contacted soccer kick.

68

External Forces

Fig. 6-12a, b, c
Soccer head—well executed.

resulting in excessive spin, lower velocity, and a misdirected kick. The force of impact will further extend the ankle joint and thus reduce the impact force, because the top of the foot is not aligned with the lower leg.

The study of the judo backfall by Deusinger (F-9) shows how a body part (arms) generates as much velocity as possible before impact, and how, with split-second timing, the body parts involved are "set" as rigidly as possible in anticipation of the impact. This timing can be identified by observing the absolute angular velocity and acceleration of the lower arms. The velocity should be at its maximum close to impact even though the deceleration of the segment begins just before impact. This indicates the stabilizing of the joints for the impact. This rapid deceleration of the arms just prior to impact also increases the downward force, but shoulder rigidity at impact is more essential. Therefore, in a well-executed judo fall the arms decelerate rapidly just before impact and hit before the trunk impacts, thus contacting the ground with as much joint stabilization as possible.

THE SOCCER HEAD

When heading a soccer ball, the motion of the trunk and head vary according to the speed of the oncoming ball. After the ball is struck and the pressure of the impact released, a certain amount of time is required to readjust to the sudden reduction of the impact force. Figure 6-12a–c shows that much more trunk motion takes place after impact than before impact, while Fig. 6-13a and b illustrates this adjustment for different ball approach velocities. The oncoming velocity of the ball is only 22 mph in Fig. 6-13a, so angular motion of the trunk about the hips is necessary to increase the striking momentum. The outgoing velocity obtained was 28.4 mph. In Fig. 6-13b the oncoming ball velocity is 26 mph, so firm contact with no trunk rotation is sufficient to obtain an outgoing velocity of 40 mph. Relative head motion was influenced by trunk motion, since there was very little motion of the head relative to the trunk when angular motion did exist. There was also a greater relative head motion when there was no angular trunk motion, although it occurred mainly after impact. Mawdsley (I-56) analyzed soccer heading and showed the dominance of the trunk and the necessity for rigidity at impact. He also found that the hip joint followed the path of the total body center of gravity more closely than any other joint; therefore, he used the hip joint as the fixed point for an analysis of a body in flight.

Anticipation of impact also presents interesting results when the impact does not occur. Figure 6-14a shows a ball being struck, with the resulting slowdown of the racket due to impact. Figure 6-14b shows the same motion, but the ball is missed inadvertently. The maximum velocity reached is the same, but this velocity continues after the anticipated contact point to point 5. The muscles react to the new situation and the movement is decelerated. This velocity change to zero takes place in a shorter distance, however, as the muscle requirement for the deceleration is much greater than when the ball is hit. If this happened when kicking, the lower leg would be moving toward its limits of extension and a knee injury or a "pulled" hamstring could result. When a batter misses a ball, the motion can easily be increased with a longer follow through, and greater muscle forces are avoided. In most sports the slow, rather than abrupt, changes in velocity on the follow through are advantageous in limiting the muscle force requirements.

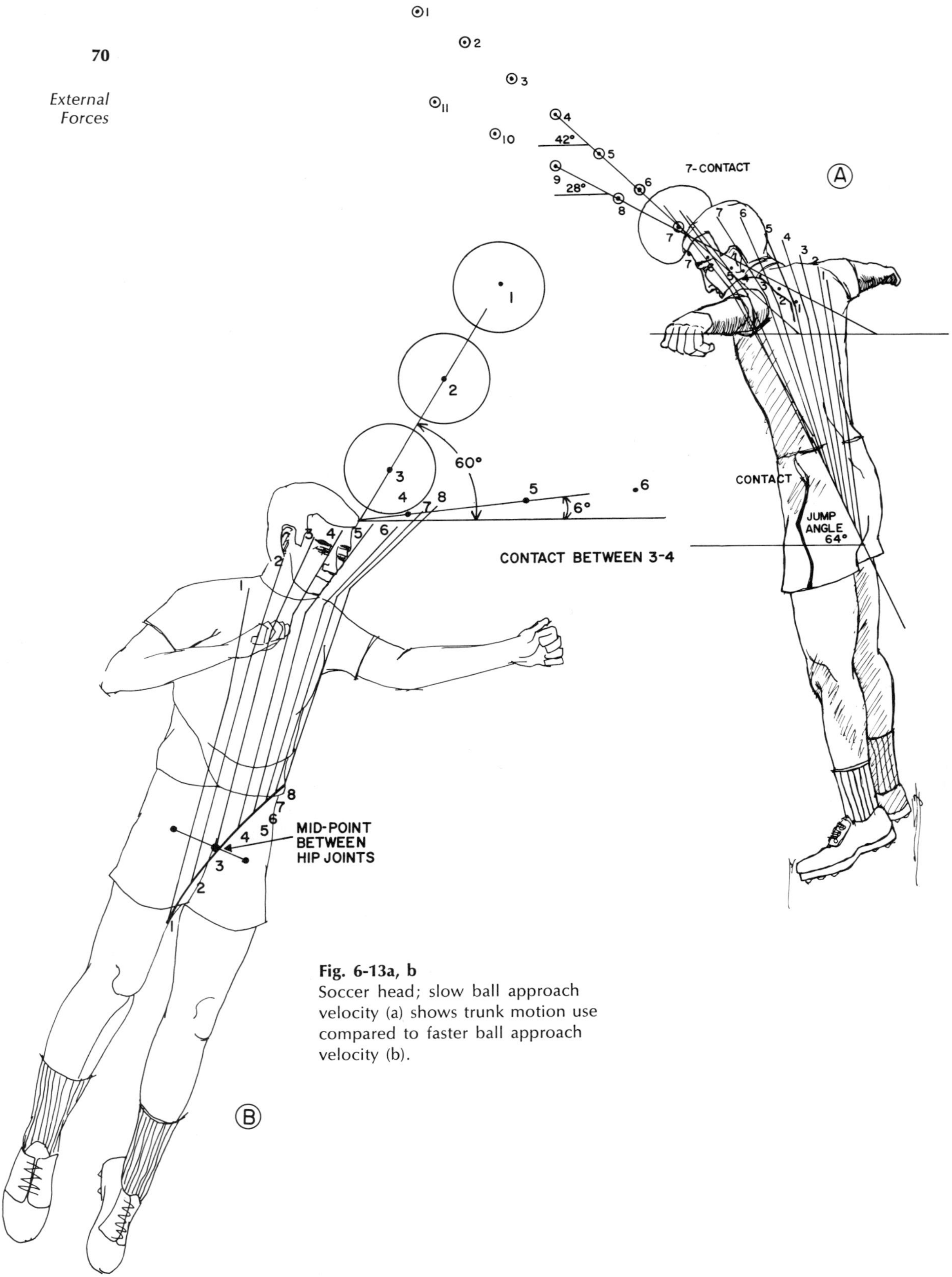

Fig. 6-13a, b
Soccer head; slow ball approach velocity (a) shows trunk motion use compared to faster ball approach velocity (b).

External Forces

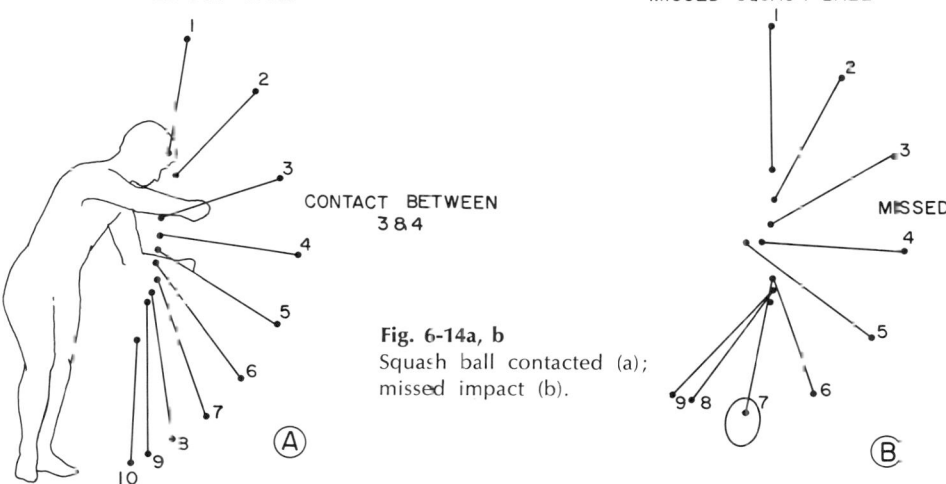

Fig. 6-14a, b
Squash ball contacted (a); missed impact (b).

IMPACT AND FOOTBALL INJURIES

An understanding of impact should be helpful in reducing injuries in body contact sports as well as in increasing the proficiency of performance. Because of the interrelationships between strengths and weaknesses of body positions and forces of impact, injuries in football are presented here to relate impact and body positions to actual injuries as filmed in games. The drawings in this section were made from blow-ups of selected frames of motion pictures to help show why the injuries occurred. The data presented below were made possible through the cooperation of the University of Massachusetts Photo Center; surgeon Dr. George Snook; trainer James Laughnane; and artist John McLaughlin.

Figure 6-15 shows a block about to be thrown across the lower leg. This block resulted in a fracture of the tibia due to the angle of the leg and the driving of the foot into the ground by the blocker. Figure 6-16 shows a player's sharp change of direction while running. This resulted in a break of the tip of the fibula because of the force of the body weight on a poorly aligned joint.

Fig. 6-15
Fracture of tibia.

External Forces

Fig. 6-16
Fracture of the tip of the fibula.

Fig. 6-17
Torn medial meniscus of the knee.

Fig. 6-18
Clip-torn medial collateral ligament.

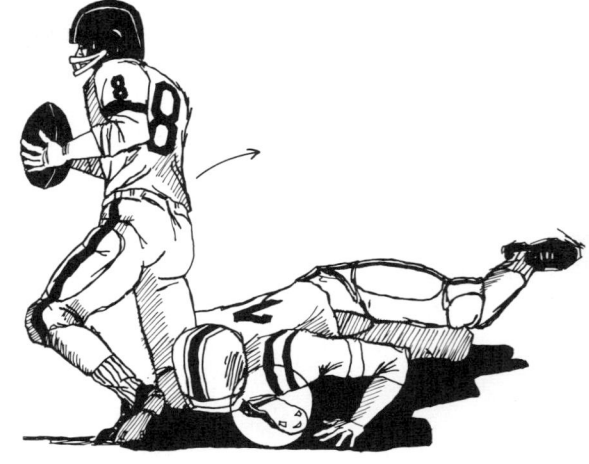

Fig. 6-19
Twisting of the knee — torn medial collateral ligament, torn anterior and posterior cruciate ligaments, torn medial meniscus.

Figure 6-17 illustrates one of the most common knee injuries, a torn medial meniscus. The tackler is in a poor initial support position with the weight on the inside of the foot. The runner forced the tackler backwards, destroying the alignment of the thigh and shank, thus placing undue stress on the medial side of the knee. The angle of the lower leg also resulted in the fixing of the foot, because a part of the impact force drove the foot into the ground. The clip shown in Fig. 6-18 also fixed the foot to the ground and resulted in a torn medial collateral ligament which required surgery. The forward motion of the runner was stopped suddenly in his own footstep and the separating force at the knee joint easily exceeded the anatomical strength.

The knee is especially vulnerable to twisting — Fig. 6-19 illustrates the most severe of such knee injuries. (This is identical to the torque injuries in skiing when a ski tip is caught.) The foot was fixed by a tackler from behind, and the runner's body was forced to his right by another tackler who bounced off. This resulted in a torn medial collateral ligament, torn anterior and posterior cruciate ligaments, torn medial meniscus, and complete loss of integrity of the joint capsule. The lower leg was actually twisted out of the socket. The lateral ligament injury is the most infrequent because it is stronger than the other ligaments of the knee, and because there is far less of a chance of receiving a blow that will open up the lateral side of the joint. Figure 6-20 shows an injury to the lateral meniscus when a defensive lineman was double teamed. One blocker hit the right leg, straightening the knee and anchoring the foot. Another blocker hit high on the left side causing a twisting and bending motion of the trunk. This placed a stretch on the medial ligament, which held, but the pinching of the lateral side in addition to the rotation of the thigh on the lower leg resulted in a torn lateral meniscus.

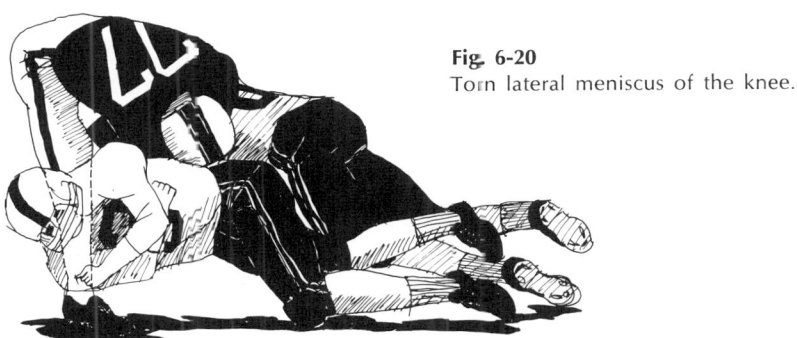

Fig. 6-20
Torn lateral meniscus of the knee.

Figure 6-21 shows a low tackle with contact directly on the knee, where no injury occurred. The insufficient holding force of the cleats, the angle of the leg, and the body position allowed the left foot to be easily displaced by a force within the tolerance of the joint structure. Figure 6-22 shows a similar impact on the outside of the leg which resulted in incapacitation but no surgery. A would-be tackler is being blocked on his blind side by a player running at full speed. A unique situation exists because the injured player lost his shoe while running, and was hit almost exactly at a time when his full weight came down on the shoeless foot. During the moment just after contact the foot

Fig. 6-21
Foot displaced—no injury.

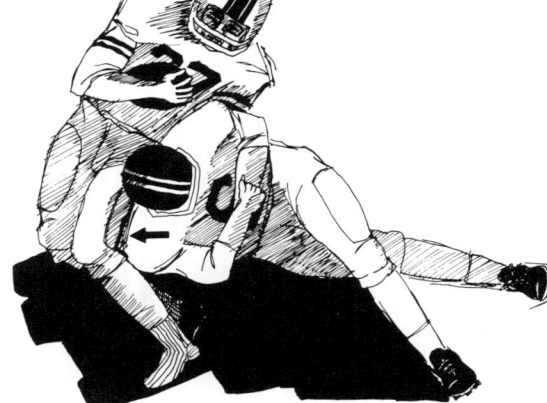

Fig. 6-22
Shoeless foot not displaced until after extreme hip rotation took place. Subluxation of the knee, no surgery.

remained fixed on the ground. The force of impact caused medial rotation of the leg at the hip joint, with the knee moving inward while the foot and head remained in place. The first movement of the foot was rotation about the heel before both feet were wiped out from under the body. This devastating block produced forces that were near the limits of human tolerance, and was diagnosed as a subluxation of the knee.

The situation in Fig. 6-22 leads to two conclusions:

1. If shoe cleats had prevented this foot rotation, the knee could not have been moved inward as far, and more serious injury could have occurred.
2. The weight on the foot is a major factor in stabilizing it, even without cleats.

The weight on the foot can be less or greater than body weight. It can be greater due to the forces of the runner's motion or due to the external forces created by the angle of contact of another player (see Fig. 6-23).

Figure 6-24 shows what can happen when the body is not fixed, but is completely upended. This is used merely to illustrate that injuries occur when the total force is concentrated on a small area. Hitting the tip of the shoulder can result in a broken collar bone, an acromio-clavicular separation, or a glenohumeral dislocation. If most of the force of impact had been on the head,

75

External Forces

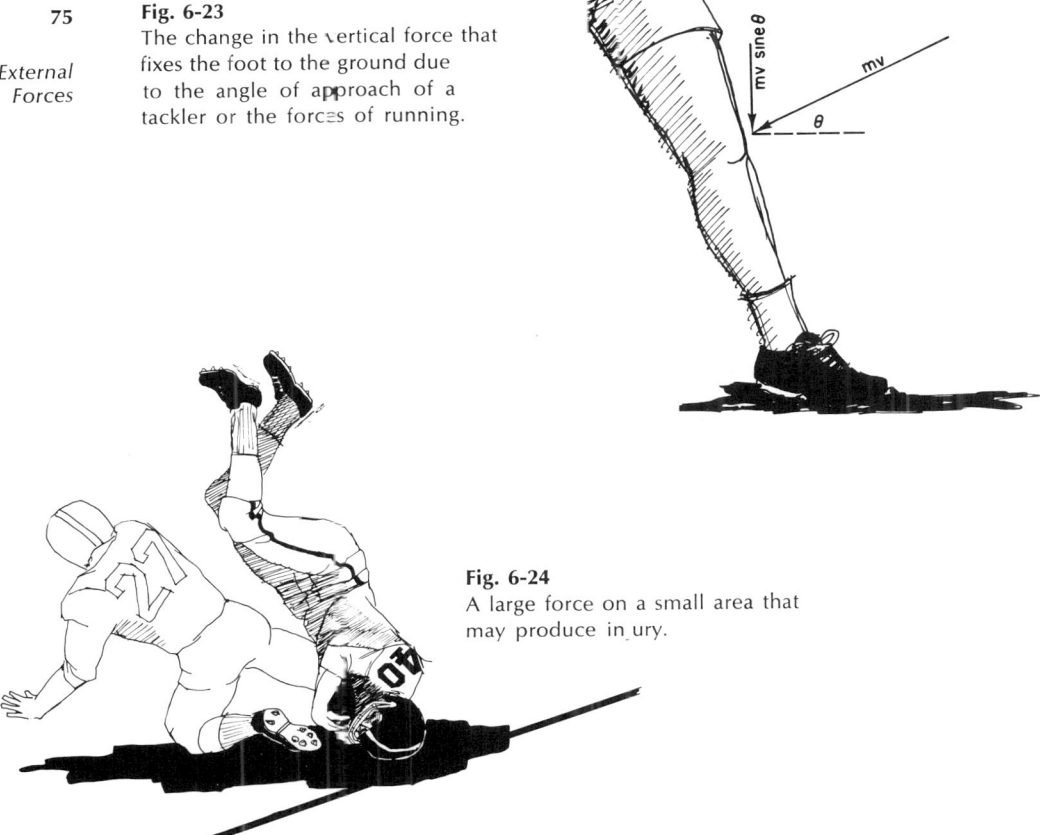

Fig. 6-23
The change in the vertical force that fixes the foot to the ground due to the angle of approach of a tackler or the forces of running.

Fig. 6-24
A large force on a small area that may produce injury.

a serious neck injury could have resulted. It is this type of situation that requires relaxation so the body can more easily collapse, thus reducing the force of impact, as well as redistribute the area of contact. The necessity for both rigidity and relaxation is again evident in the judo backfall. The arms and trunk muscles are rigid, while the legs are relaxed so that their continued motion allows a slower deceleration, thus reducing the force of impact.

This is a small sample of documented injuries, but it can be helpful in studying injury prevention. It is evident that fixing the foot must be avoided. Short steps that eliminate the vulnerable angle position of the leg should be practiced. The proper placement of the foot while running (pointing the toes in the direction of the movement) will keep the weight off the inside edge of the foot (Fig. 6-25). If a toe-out position of an angled leg is contacted, the medial ligament sustains the brunt of the force, and is thus very liable to injury.

The strongest anatomical positions should be preserved at all times. This means maintaining proper alignment of the body segments. The thigh, shank, and foot should remain in the same plane, and the vertebral column should be kept in good alignment. The elbows should be kept lower than the shoulders when possible, since the glenohumeral joint is more susceptible to injury when the upper arm is raised. The body's ability to absorb shock with the muscles rather than cause undue joint stress is all a matter of proper body positioning.

External Forces

Fig. 6-25
Proper foot alignment when changing direction to avoid knee injury.

However, extreme rotation of any joint should be avoided, as should absorption of blows when muscles are fully stretched, near their extreme range of motion. Muscle tears will occur when a stress is placed on a fully extended muscle (Fig. 6-26), and joint stress occurs when a force is placed on a joint already at the limits of its range of motion.

Fig. 6-26
Tearing of the adductors of the right leg due to a stress placed on the nearly fully extended muscle.

chapter 7

SPORTS EQUIPMENT— IMPACT—BALL SPIN

MOTION ANALYSIS OF SWING AND IMPACT

A racket, golf club, or bat presents an additional problem in the analysis of human motion, both during the swing and during impact. The weight and pivot point of the combined hand (or hands) and implement, and the common center of gravity and radius of gyration, must be determined to obtain the correct data for analysis of the last segment in the link system. (See Fig. 4-17.) All of these measurements will differ during impact from those during the swing. The data for a tennis, squash, and badminton racket, a baseball bat, and golf clubs are presented in Table 7-1 and illustrated in Fig. 7-1.

The pivot point during a swing is the right wrist joint when one hand is involved and the left wrist joint when two hands are involved (as in golf). However, during impact the racket, club, or bat is attempting to rotate about the base of the right index finger, with the other fingers countering the torque caused by the impact (Fig. 7-2). When both hands are involved, the left hand aids the fingers of the right hand in countering the torque of impact. The rackets, clubs, and bat were swung about the handle end with no weight added, and with weight added to represent the hand or hands; and about the pivot point (location of base of index finger on the handle) during impact. This was done to obtain the period of oscillation to calculate the moment of inertia (I_o).

Sports Equipment — Impact — Ball Spin

BADMINTON RACKET 67.5 cm

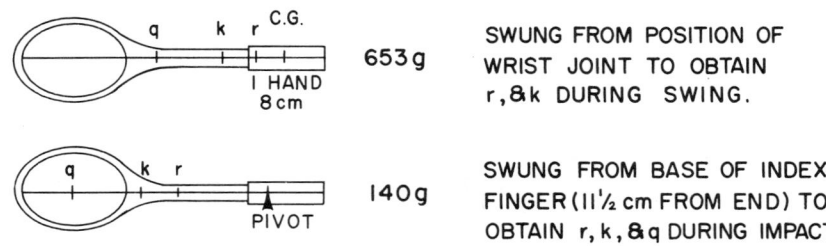

653 g — SWUNG FROM POSITION OF WRIST JOINT TO OBTAIN r, & k DURING SWING.

140 g — SWUNG FROM BASE OF INDEX FINGER (11½ cm FROM END) TO OBTAIN r, k, & q DURING IMPACT

GOLF CLUBS

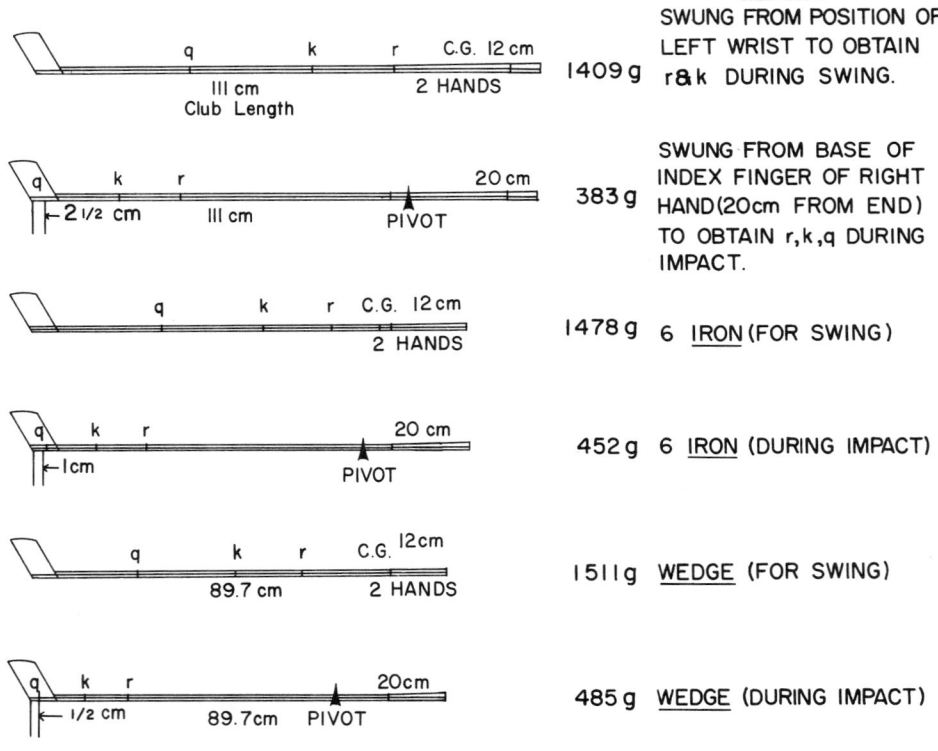

1409 g — **DRIVER** SWUNG FROM POSITION OF LEFT WRIST TO OBTAIN r & k DURING SWING.

383 g — SWUNG FROM BASE OF INDEX FINGER OF RIGHT HAND (20 cm FROM END) TO OBTAIN r, k, q DURING IMPACT.

1478 g — 6 IRON (FOR SWING)

452 g — 6 IRON (DURING IMPACT)

1511 g — WEDGE (FOR SWING)

485 g — WEDGE (DURING IMPACT)

Fig. 7-1
The center of gravity (r), radius of gyration (k), and center of percussion (q) of selected sports equipment.

Sports Equipment — Impact — Ball Spin

BASEBALL BAT 82.5 cm

925 g — SWUNG FROM END.

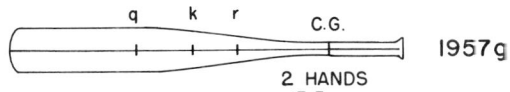

1957 g — SWUNG FROM POSITION OF LEFT WRIST JOINT TO OBTAIN r & k DURING SWING

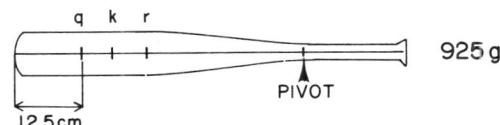

925 g — SWUNG FROM BASE OF RIGHT INDEX FINGER (20 cm FROM END) TO OBTAIN r, k, & q DURING IMPACT.

TENNIS RACKET 67.5 cm

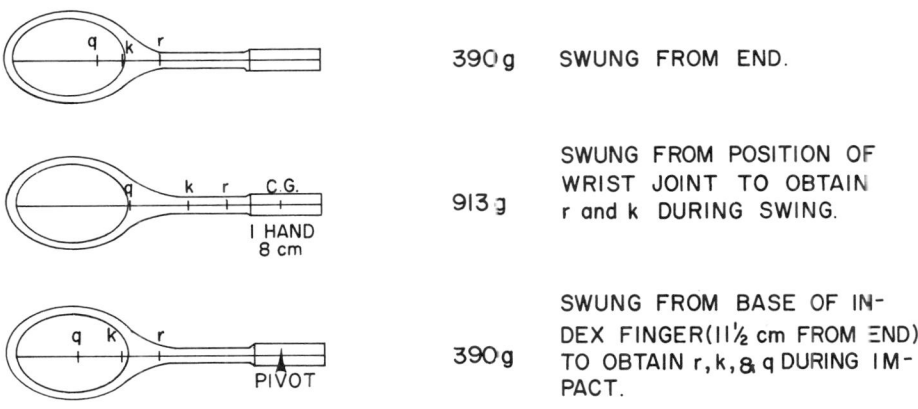

390 g — SWUNG FROM END.

913 g — SWUNG FROM POSITION OF WRIST JOINT TO OBTAIN r and k DURING SWING.

390 g — SWUNG FROM BASE OF INDEX FINGER (11½ cm FROM END) TO OBTAIN r, k, & q DURING IMPACT.

SQUASH RACKET 67.5 cm

810 g — SWUNG FROM POSITION OF WRIST JOINT TO OBTAIN r and k DURING SWING.

295 g — SWUNG FROM BASE OF INDEX FINGER (11½ cm FROM END) TO OBTAIN r, k, & q DURING IMPACT.

Sports Equipment — Impact — Ball Spin

Table 7-1 Swing and Impact Data on Sports Equipment

Equipment	C.G. of Hands (cm)	Pivot Point (cm)	Weight (gm)	r (cm)	k (cm)	q (cm)	l (cm)	T (sec)	$I_o (= mk^2)$ (gm-cm²)
Baseball Bat		End	925	55.2	59.6	64.4	82.5	1.61	3350
Swing	13	End	1957	33.1	42.6	55	82.5	1.49	3640
Impact		20	925	34.2	41.1	49.4	82.5	1.41	1595
Badminton Racket		End	140	31.4	38.4	46.8	67.5	1.37	210
Swing	8	End	653	13	21.4	35.3	67.5	1.19	306
Impact		11½	140	20	29.6	43.8	67.5	1.33	125
Golf Putter		End	475	67.5	73.9	80.7	87.6	1.8	5450
Wedge		End	485	69.1	76.5	84.2	89.4	1.84	2875
Swing	12	End	1511	30.5	45	66.5	89.7	1.63	3120
Impact		20	485	49.1	58.5	69.2	89.4	1.67	1680
9 Iron			471	69.5	75.5	82.6	89.3	1.82	2730
8			457	69.6	77.2	85.4	90.2	1.85	2770
7			462	69.4	76.9	85	91.6	1.85	2780
6		End	452	70.2	77.2	84.8	93.1	1.85	2750
Swing	12	End	1478	30	45	67	93.3	1.64	2165
Impact		20	452	50.6	61.1	73.4	93.1	1.72	1690
5			445	70.6	78	86	94.5	1.86	2760
4			440	70.9	79	88.2	95.8	1.88	2810
3		End	427	72.4	80.5	89.8	97	1.9	2840
Swing	12	End	1453	30	44	64.5	97	1.65	3010
2			427	72.5	81	89.7	98.5	1.9	2850
4 Wood			400	76	85.2	95.4	106.6	1.96	2960
3			387	76.8	86.6	97.7	108.0	1.98	2960
2			383	76.9	87.5	99.7	109.5	2	2995
Driver		End	383	77.7	87.6	99	111	2	3020
Swing	12	End	1409	30.5	46	76.4	111	1.75	3340
Impact		20	383	57.7	71.6	88.8	111	1.89	2000
Squash		End	295	34	40.7	48.7	67.5	1.4	503
Swing	8	End	810	17.5	26.8	40.9	67.5	1.28	591
Impact		11½	295	22.5	32.2	46	67.5	1.36	314
Tennis Child's Racket		End	345	33.3	38.6	44.8	65	1.34	525
Medium		End	390	34.4	39.8	46.1	67.5	1.36	630
Heavy		End	410	34.3	39.8	46	67.5	1.36	644
Swing (Medium)	8	End	913	19.5	28.2	40.7	67.5	1.28	742
Impact		8	390	23	31.2	41.9	67.5	1.3	384
75 gm weight on tip		End	475	39.5	44.7	50.8	67.5	1.43	970

$$I_o = \frac{T^2 wr}{4\pi^2} \qquad k = \sqrt{\frac{I}{m}} \qquad q = \frac{k^2}{r}$$

If the location of the hands is unchanged, the center of percussion then depends on the location of the center of gravity of the implements. A change in the position of the hands on a given implement will also change the center of percussion because the pivot point during contact has been changed. Thus both the structure of the equipment and the hand placement affect the center of percussion.

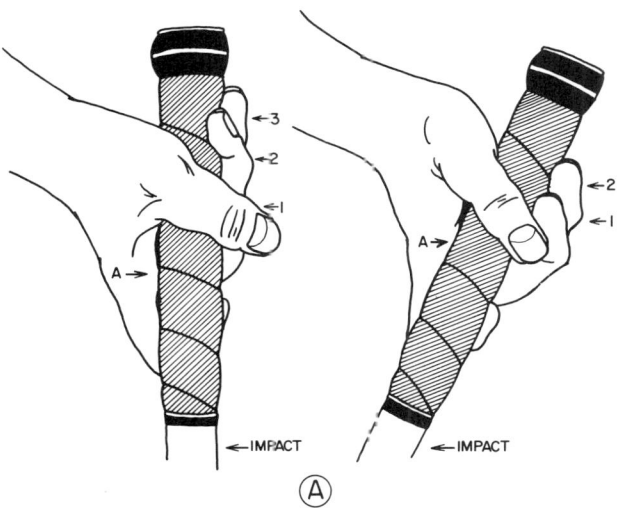

Fig. 7-2a
Torque forces during impact of a tennis ball. The fingers (1, 2, 3) counter the force of impact about the pivot point (A).

Fig. 7-2b
Torque forces during impact of a golf ball. The left hand counters the force of impact about the pivot point of the right hand.

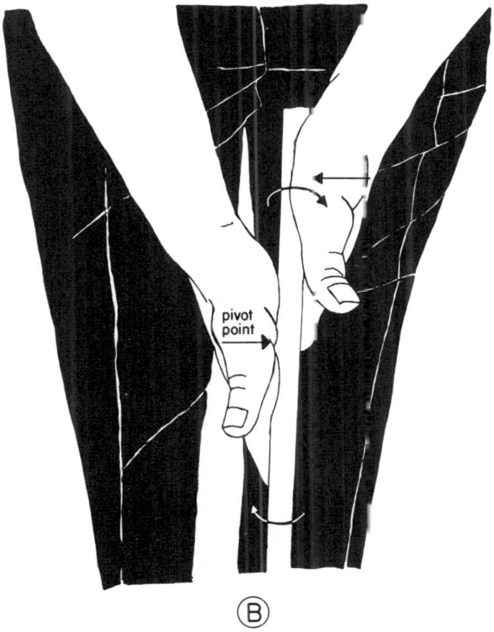

THE PROPERTIES OF BALLS

Friction

Using sports equipment means hitting, catching, or throwing a ball, so naturally the properties of balls must be considered. The coefficient of friction is well covered in almost any physics book and Rabinowicz (C-35, C-36) explains frictional variations fully. (See also Problems 6, 11, and 12 in Appendix A).

Static friction or friction during very slow movement is greater than friction during fast motion. The difference between static and sliding friction is well illustrated by Tricker and Tricker (H-27): they block the back wheels of a toy car and run it down an incline. The sliding friction of the back wheels is less than the static friction of the front wheels, and thus the car turns around and goes down backwards. It must be understood that static friction exists between the front wheels and the incline, as all relative motion allowing motion takes place at the axle.

The definition of friction is complicated, but application of the laws of friction is relatively simple. The following laws should be helpful when applied to various sports.

Laws of Friction

1. The coefficient of friction μ is the relation between a force F necessary to move a body along a plane at a uniform rate, and a normal force N perpendicular to that plane (the plane may be horizontal or inclined; see Problem 6 in Appendix A):

$$\mu = \frac{F}{N}$$

N equals the weight on a horizontal surface; as the weight of an object is increased the force necessary to move the object increases so that the coefficient of friction is unchanged. Friction, therefore, is not affected by an object's weight.
2. The coefficient of friction is independent of the contact area of the surfaces.
3. Although Rabinowicz shows that the coefficient of sliding friction decreases with velocity, for practical use in sports, friction may be considered to be unaffected by velocity.
4. Since the coefficient of friction varies with the properties of the contact surface, any comparison of balls must be done on the same surface.

Coefficient of Restitution

The coefficient of restitution (e) of a ball bounced on a specified surface is obtained by dividing the ball velocity after impact by the ball velocity before impact:

$$e = \frac{\text{velocity after impact}}{\text{velocity before impact}}$$

If the equipment for measurement of ball velocity is not available, the coefficient may be determined by dropping the ball from a given height. An increase in drop height increases the ball velocity before impact, and by measuring the rebound height, the coefficient may be determined:

$$e = \sqrt{\frac{\text{rebound height}}{\text{drop height}}}$$

If a ball is hit or dropped against a surface, the coefficient decreases as the velocity before impact increases. In a comparison between various balls (see Table 7-2), a firm, wood floor was used to obtain both the coefficient of friction and the coefficient of restitution. The balls were first dropped from 100 inches and then kicked or thrown to obtain velocities between 50 and 60 mph. Remember that these data differ substantially under other conditions. Weights and diameters of balls and weights of other equipment are included in Table 7-2 to aid in problem solving. Note that as the velocity increases, soft balls produce a greater change in the coefficient of restitution than do harder balls.

Table 7-2   Properties of Balls

BALL	DIAMETER (cm)	WEIGHT (gm = lb)		COEFFICIENT OF RESTITUTION ON WOOD		COEFFICIENT OF FRICTION ON WOOD
				Dropped	50–60 mph	
Baseball	7.3	149 =	.324	.5	.44	.34
Basketball	24.5	600 =	1.32	.75	.64	.28
Golfball	4.4	47 =	.104	.6	.58 .75 (metal)	.2
Handball	4.8	61 =	.134	.8	.5	.75
LaCrosse Ball	6.5	150 =	.331	.7	.6	.5
Paddleball	6.0	42 =	.093	.7	.45	.6
Ping Pong Ball	4.0	3 =	.0056	.8	.7	.25
Soccer Ball	22.0	425 =	.935	.75	.65	.3
Squash Ball	4.5	32 =	.705	.52	.4	.45
Softball	10.3	171 =	.377	.55	.4	.33
Superball	5.0	63 =	.139	.9	.85	.7
Tennis Ball	6.8	58 =	.128	.74	.52	.25
Volleyball	20.5	300 =	.66	.75	.68	.3
Other Equipment						
Badminton, shuttlecock		5 =	.011			
Discus		1620 =	3.56			
Football		415 =	.915	.7		.23
Hockey Puck		167 =	.368			
Javelin		800 =	1.755			

APPLICATION OF THE LAWS OF FRICTION AND RESTITUTION TO SPORTS

Tennis

The game of tennis varies considerably depending on the court surface and the type of ball used. As ball velocity increases, the coefficient of restitution of a nonpressure ball decreases more rapidly than that of a pressure ball. Therefore, when playing with a nonpressure ball the hard hitter is at a disadvantage with regard to ball velocity. Court surfaces vary so greatly that a certain player can excel on one surface and be mediocre on another. A ball that leaves a server's racket at 115 mph is going about 85 mph as it hits the service court, where ground friction then further reduces the ball velocity. By the time the ball reaches the receiver's racket, the ball velocity may be 50 to 65 mph, depending on the court surface. Court variations which affect both friction and restitution can be measured only under very controlled conditions. Other variables that must be accounted for are: ball spin, wind, ball velocity, and the angle of the ball's approach to the ground.

Squash

The game of squash also presents problems due to variations in friction and restitution. When the first United States Open Tournament was held in 1955, the ball used bounced 34 to 36% of its drop height when dropped from 100 inches. This favored the hard hitter and made it difficult to make the more delicate shots of the game. It was recognized that a ball that rebounded 25 to 31% of a drop height of 100 inches would be the best for all shots to be used. The weight and compressibility of the balls were also standardized as much as possible. Even so, balls had unusual bounces under certain conditions for unknown reasons; this resulted in unexpected errors. It was found that the coefficient of friction between the rubber ball and a floor paint with grit added was very high, so that a ball hit with a great deal of spin had a very exaggerated bounce. Efforts by ball manufacturers and a change in floor specifications have helped to standardize playing conditions.

Soccer

A variation in the coefficient of restitution of soccer balls once influenced the skills of passing, trapping, and kicking. The rules now standardize the balls so the limits of the coefficient of restitution are .7 to .74 when dropped from 100 inches. The control of the ball's weight when wet has also been a major improvement. A few supposedly waterproof balls gained over 100 grams when wet, so the total weight was as high as 540 grams. Restriction of the total weight to 420 grams, even when wet, has greatly reduced the force of impact, especially for heading. Eye injuries, concussions, and even death have resulted from heading a soccer ball.

Volleyball

Volleyball has been faced with the same wide variation in the coefficient of restitution for balls that had the pressure specified. Several manufacturers sent balls to the University of Massachusetts Bio-Mechanics Lab at the request of the U.S. Volleyball Association. Of ten balls tested, only two met the rule specifications. The manufacturers were made aware of the inadequacies and now attempt to meet the requirements for weight, size, ball pressure, and a coefficient of restitution producing a rebound of 60 to 65 inches from a 100 inch drop.

BALL SPIN

When a ball is contacted by a club, racket, or bat, the laws of friction, restitution, and momentum are all involved in the determination of the ball velocity, direction, and spin after impact. The momentum of each colliding object must be used, and the final ball direction, velocity, and spin must be calculated from the motions of the ball and the striking implement.

The angle of the face of the racket or club determines the direction of the ball due to the momentum of the implement — this direction is called the *line of action*. The line of action for two colliding balls is drawn by connecting the ball centers. The direction of swing of the implement relative to the line of action determines the spin placed on the ball, and determines the deviation of ball flight from the line of action due to the friction of impact. To help understand how ball spin is attained, Fig. 7-3a to g presents the results of impact between a tennis racket and ball.

The most simplified conditions are assumed in Fig. 7-3a through g to illustrate how ball spin results from the angle of the swing relative to the angle of the racket face. The masses are considered equal, the ball has no spin before impact, and the coefficient of restitution of 1 makes the ball rebound angle equal to the approach angle, if friction is neglected. Fig. 7-3a has the racket face, racket swing, and ball flight before impact all in line. Fig. 7-3b changes only the direction of the incoming ball. Fig. 7-3c shows the racket face tilted 25° to the direction of the swing, and isolates the other factors by striking a suspended ball. Fig. 7-3d duplicates Fig. 7-3c, but has the ball approaching from a direction opposite to the swing direction. Fig. 7-3e duplicates Fig. 7-3c, but has the ball approach from above rather than straight on. Fig. 7-3f keeps the racket face perpendicular, but now changes the direction of the swing. Again, a suspended ball is hit to isolate the resultants of the swing (velocity and spin). Fig. 7-3g duplicates Fig. 7-3f, but has the ball moving upward before impact. Fig. 7-3h includes the masses of the colliding ball and racket, and obtains the ball direction after impact vectorally. Ball velocity still cannot be determined unless the change in racket velocity is known after impact; the striking mass may then be determined.

BALL REBOUND ON A SOLID SURFACE

The complexity of determining ball direction and velocity increases when the coefficient of restitution, coefficient of friction, and ball spin before impact are considered. Professor Corrado Poli of the Mechanical and Aerospace Engineering Department, University of Massachusetts, determined the formulas necessary to calculate the final velocity (V_f), spin (ω_f), and angle of rebound (θ_f), given the initial velocity (V_i), spin (ω_i), approach angle (θ_i), coefficient of friction (static, μ_s, dynamic, μ_d), coefficient of restitution (e), and ball radius (r) (see Fig. 7-4). The formulas for both a slip and no-slip condition are presented with the conditions that determine slippage. The two- and three-dimensional formulas are presented in Appendix D so problems such as the "hop" serves in tennis and handball may be solved. The two-dimensional formulas were used to calculate approximately 100 combinations of the five variables. The conclusions are as follows:

1. The angle of rebound (θ_f) is greater than the approach angle (θ_i) when only the effect of friction is considered (see Fig. 7-4a). This remains true regardless of the ball velocity (V_i). The angle change is 14° greater when $\theta_i = 45°$, 7.5° greater when $\theta_i = 70°$, and 11° greater when $\theta_i = 20°$. The percentage change in velocity due to friction remains constant for a given angle as V_i is varied, but the percentage increases as the angle (θ_i) decreases. (When $\theta_i = 70°$, V_f is 4% less than V_i; when $\theta_i = 45°$, V_f is 17.5% less than V_i; and when $\theta_i = 20°$, V_f is 34% less than V_i).
2. The angle of rebound (θ_f) is less than the approach angle (θ_i) when only the coefficient of restitution is considered (See Fig. 7-4a). The percentage change in velocity is constant for a given angle as V_i changes, but is larger as θ_i is increased. For a coefficient of restitution of .5: (When $\theta_i = 20°$, V_f is 7% less than V_i; when $\theta_i = 45°$, V_f is 27.5% less than V_i; when $\theta_i = 70°$, V_f is 45% less than V_i).
3. Angular velocity (ω_f, clockwise) increases due to friction proportional to the increase in V_i, but increases as θ_i decreases. For a ball with a radius 2.5 cm, ω_f increases 24.6 rad/sec when $\theta_i = 70°$, 51 rad/sec when $\theta_i = 45°$, and 67.6 rad/sec when $\theta_i = 20°$ for every increase in V_i of 300 cm/sec.
4. Backspin ($+\omega_i$) increases θ_f, and topspin ($-\omega_i$) decreases θ_f. A topspin ball produces a greater increase in velocity than backspin produces a decrease in

Sports Equipment — Impact — Ball Spin

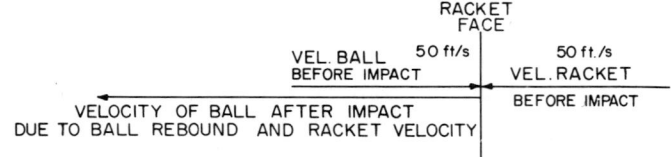

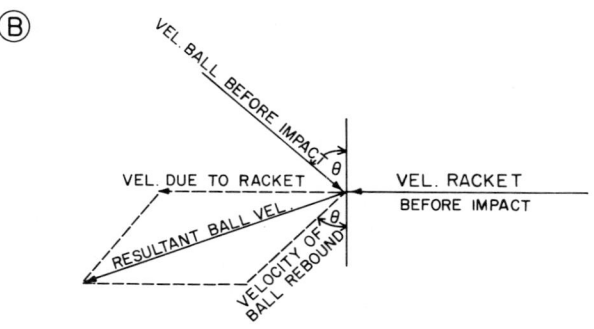

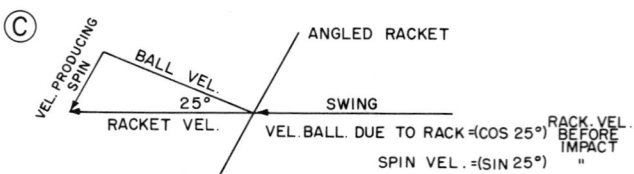

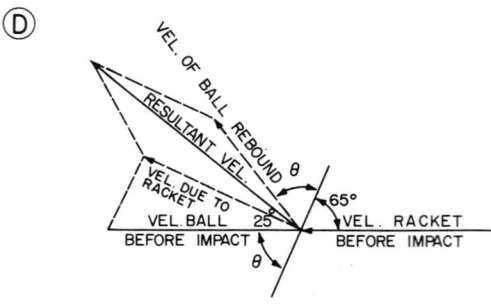

Fig. 7-3a through h
The determination of a tennis ball rebound angle and velocity under selected striking conditions.

Sports Equipment — Impact — Ball Spin

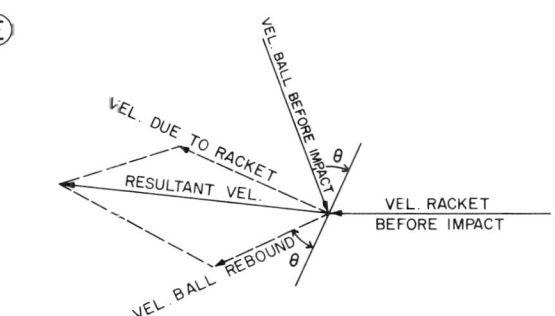

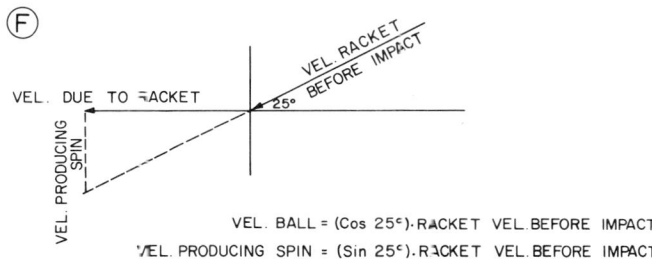

VEL. BALL = (Cos 25°)·RACKET VEL. BEFORE IMPACT
VEL. PRODUCING SPIN = (Sin 25°)·RACKET VEL. BEFORE IMPACT

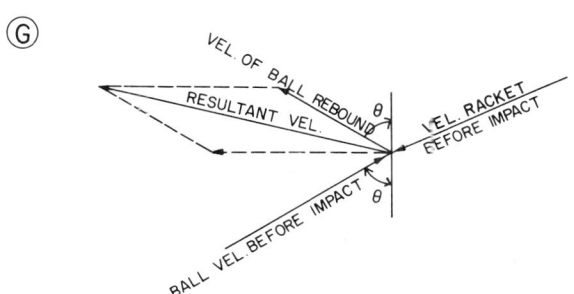

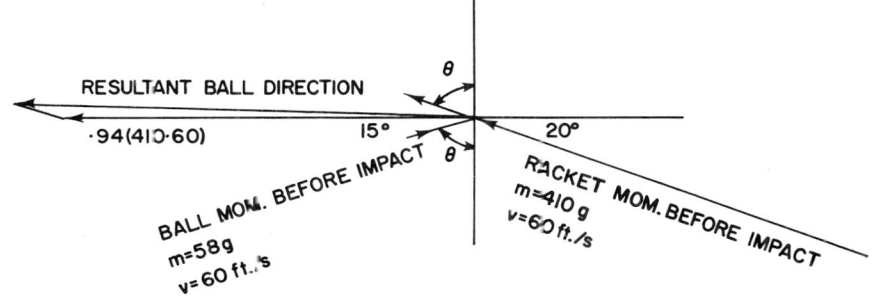

MOMENTUM CONSIDERED TO OBTAIN BALL DIRECTION AFTER IMPACT.

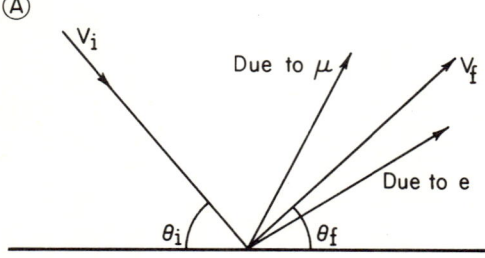

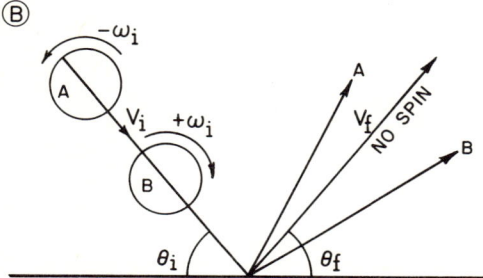

Fig. 7-4a, b
Ball rebound on a solid surface (also see Appendix D).

velocity if $e = 1$, but backspin produces a greater increase in θ_f than topspin produces a decrease in θ_f (see Fig. 7-4b).

5. ω_f decreases proportional to the increase in ball radius when all other conditions are kept constant.

Specific conditions illustrate the changes that occur due to ball spin, friction, and restitution. Given $V_i = 900$ cm/sec, $\omega_i = 0$, $\theta_i = 20°$, $e = .5$, and $r = 2.5$ cm, the rebound conditions are: $V_f = 530$ cm/sec, $\omega_f = -202$ rad/sec, and $\theta_f = 17°$. By changing only the coefficient of restitution to one ($e = 1$), θ_f equals 31°. Therefore, the effect of friction is $+11°$, and the effect of restitution is $-14°$.

By giving the ball topspin ($\omega_i = -400$ rad/sec), the data after impact becomes: $\omega_f = -363$ rad/sec, $V_f = 920$ cm/sec, and $\theta_f = 10°$, or the effect of ω_i changed θ_i by $-7°$. Giving the ball backspin ($\omega_i = +400$ rad/sec), the data after impact becomes: $\omega_f = -43$ rad/sec, $V_f = 188$ cm/sec, and $\theta_f = 55°$, or the effect of ω_i changed θ_i by $+38°$. Therefore, the change in θ_i due to friction, restitution, and ball spin would be $+11°$, $-14°$, and $-7°$, respectively, when $\omega_i = -400$ rad/sec, and $+11°$, $-14°$, and $+38°$, respectively, when $\omega_i = +400$ rad/sec.

It is evident from the above data that there is only a remote chance that θ_f will be the same as θ_i. It is also evident that a ball-floor bounce having a high coefficient of friction and a high coefficient of restitution is likely to result in a θ_f that is larger than θ_i, and that a low coefficient of friction combined with a low coefficient of restitution is likely to result in a θ_f that is smaller than θ_i. The changes are so extensive as θ_i, V_i, ω_i, e, and r are varied that specific problems should always be calculated. (References C-3, C-4, C-22, C-25, and C-40 should be consulted for ball-spin problems.)

chapter 8

THROWING AND KICKING

THROWING

In many sports an object is thrown as far or as fast as possible (baseball, javelin, discus, shot-put, hammer, and football). Other sports require that the hand reach an optimum velocity so as to impart maximum speed to a ball through impact (handball, volleyball), or to produce a maximum striking force (boxing, karate). Still other sports require accuracy of throwing rather than speed (basketball, bowling), and some of these make use of equipment (lacrosse, jai lai). A large group of sports use equipment to impart controlled maximum velocity to an object through impact (tennis, golf, badminton, squash, batting, hockey slap shot, field hockey). All of these sports have a common denominator for good performance denoted by the general heading — "throwing motion."

Throwing motion is defined as: the properly timed coordination of accelerations and decelerations of all body segments in a sequence of action from the left foot to the right hand that produces maximum absolute velocity of the right hand. (The definition applies to the equipment if a racket, club, or bat is used.) To achieve maximum hand velocity, the use of optimum positions of anatomical strength must be considered as well as the proper use of the body's link system.

The Link System

The throwing-motion link system is represented by a series of lines drawn from the left ankle joint to the knee, left hip, across to the right hip, and up to the right shoulder, elbow, and wrist (Fig. 8-1). This sequence occurs unless a running motion prevents the setting of the left foot. In the latter case, motion

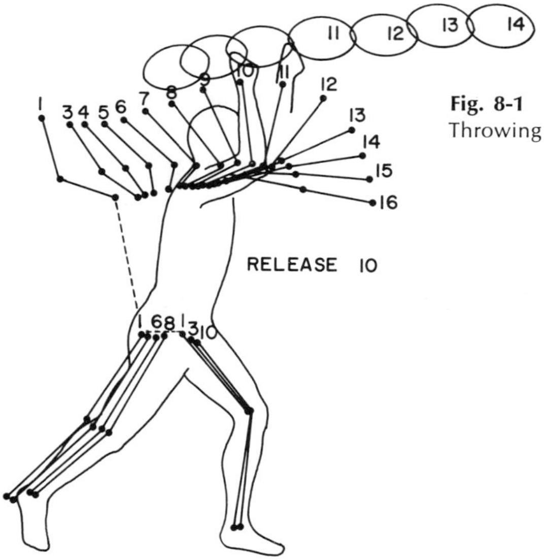

Throwing and Kicking

Fig. 8-1
Throwing (football pass) — link system.

Fig. 8-2
Badminton—jumping smash.

Fig. 8-3
Football pass for maximum distance with the front foot properly in position (Greg Landry).

takes place in the air, as in the running shot in lacrosse or the backward jumping smash in badminton or tennis (Fig. 8-2). Sometimes the running throws are desirable and sometimes unavoidable, but the function of the left leg changes from being a part of the link system to receiving the weight of the body after the motion is completed. For maximum use of the full link system, it is necessary to plant the front foot and throw with a full motion as in the sequence of Fig. 8-3, which shows Greg Landry throwing a 50-yard completion during a game. Overhand throwing motions in other sports have patterns similar to throwing a ball. The arm actions of the vollyball spike (Fig. 8-4) and badminton smash (Fig. 8-5) are similar, although the difference in the force of impact is considerable. When the left arm aids the control and velocity, as in golf, batting, and lacrosse, the sequence of action requires an analysis of both sides of the body. The throwing motion of the lacrosse shot (Fig. 8-6) shows a low right

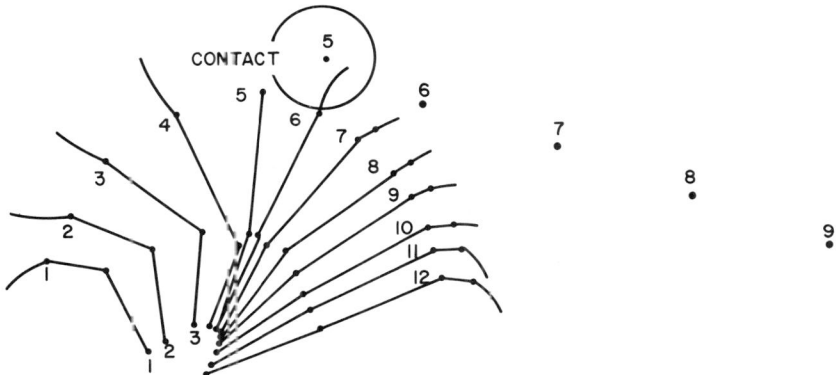

Fig. 8-4
Volleyball spike, arm action.

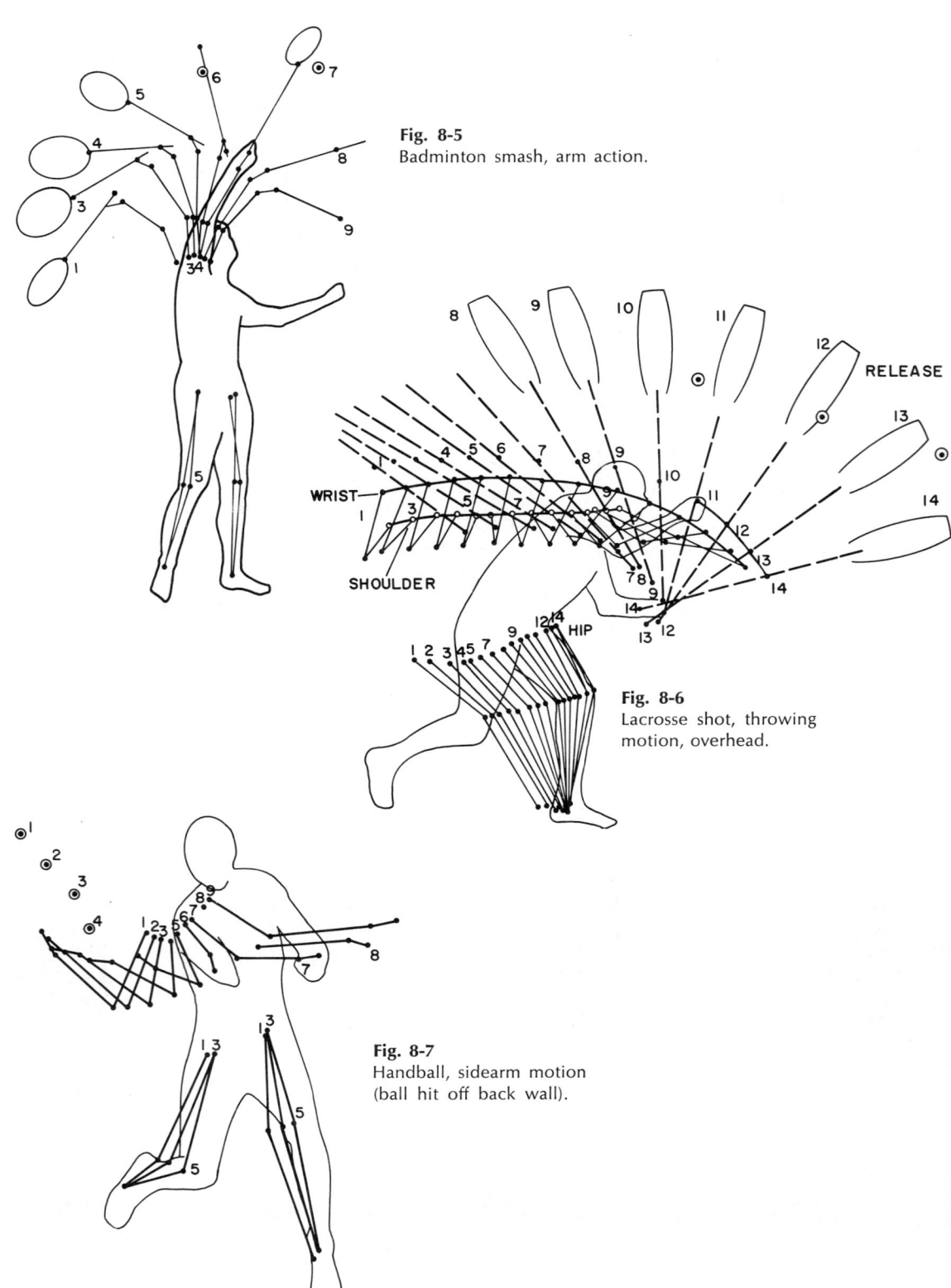

Fig. 8-5 Badminton smash, arm action.

Fig. 8-6 Lacrosse shot, throwing motion, overhead.

Fig. 8-7 Handball, sidearm motion (ball hit off back wall).

Throwing
and
Kicking

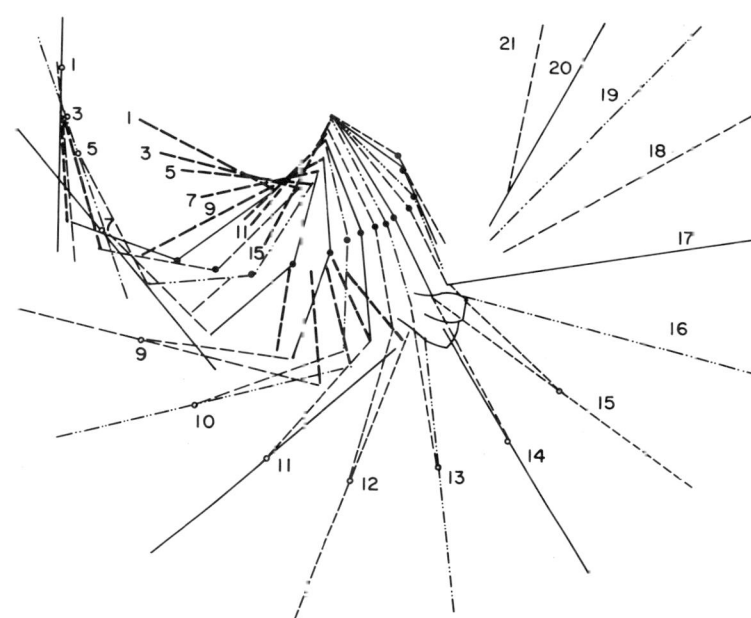

Fig. 8-8
Batting, top view.

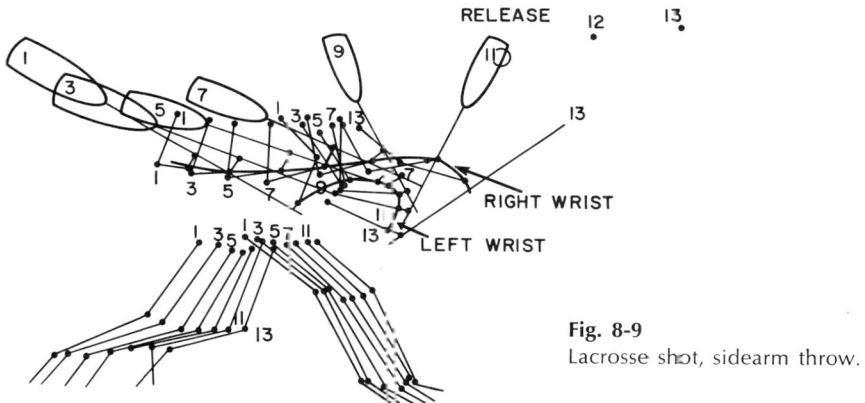

Fig. 8-9
Lacrosse shot, sidearm throw.

elbow as the left hand and stick provide the leverage necessary to aid the motion.

Sidearm motions with similar patterns exist in handball, batting, and lacrosse (Figs. 8-7, 8-8, and 8-9), and throwing motions with the arm low are used in handball and hockey (Figs. 8-10 and 8-11). The motion pattern for controlled velocity differs from that used to produce maximum velocity. Whereas maximum use of trunk rotation and a full motion are characteristic of hard throws, trunk stabilization and the restricted use of body segments are characteristic of controlled throws. Basketball and bowling are representative of the controlled-motion sports, where eye alignment and controlled velocity at release are more important than maximum release velocity (Figs. 8-12 and 8-13). Although

Throwing and Kicking

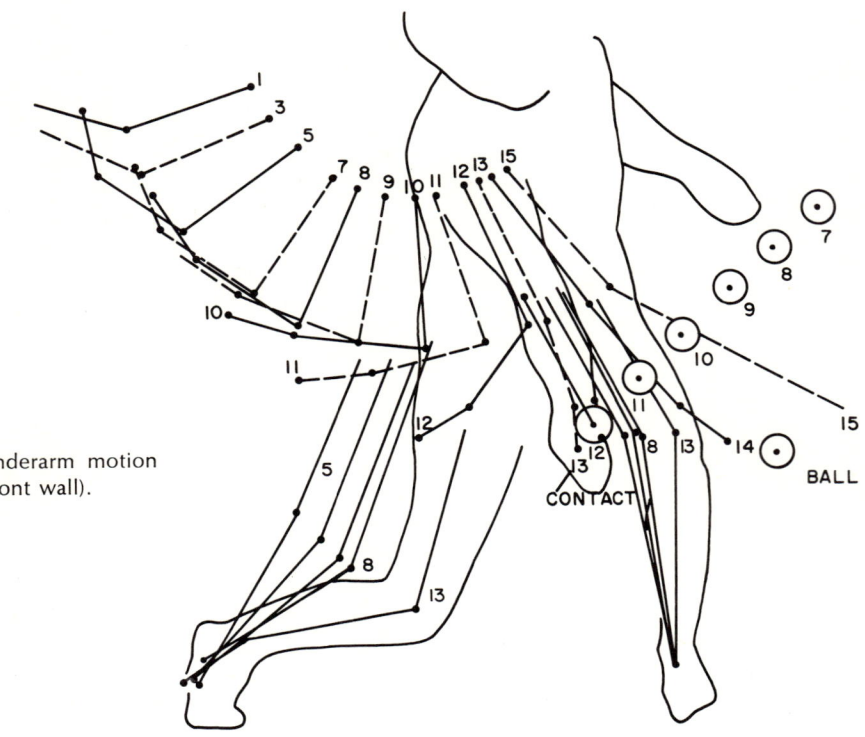

Fig. 8-10
Handball, underarm motion (ball from front wall).

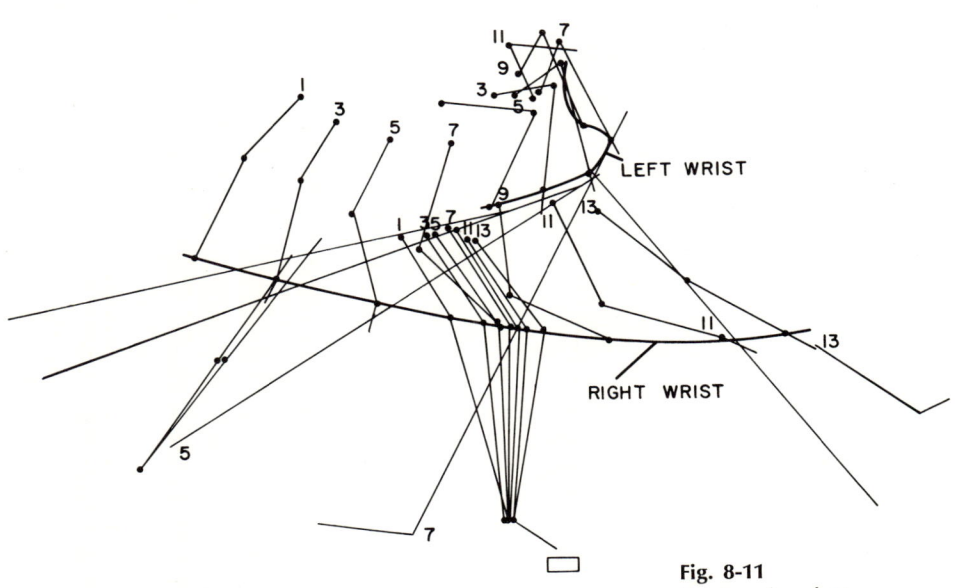

Fig. 8-11
Hockey slap shot.

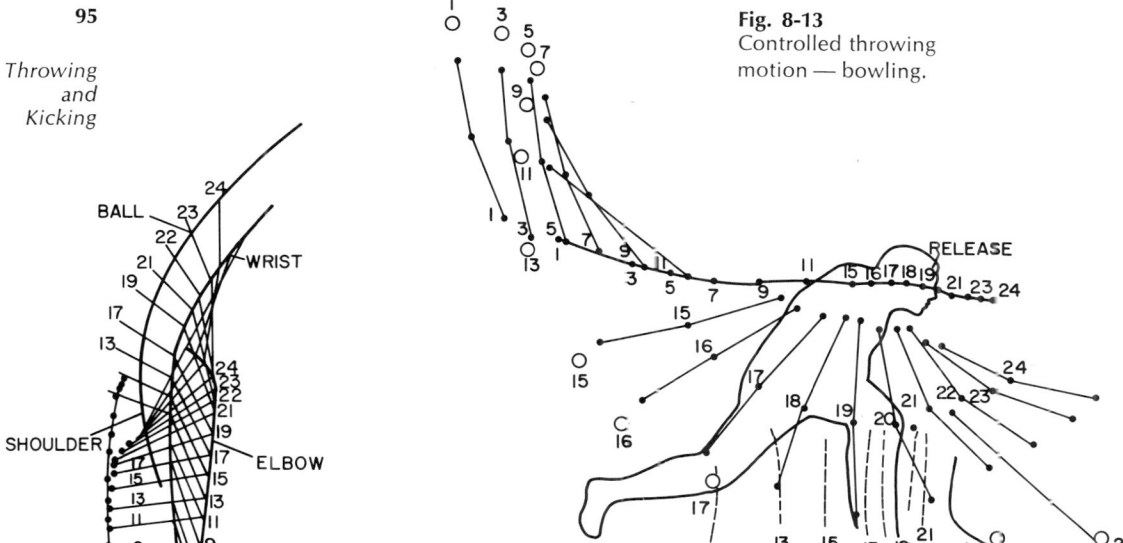

Fig. 8-13 Controlled throwing motion — bowling.

Fig. 8-12 Controlled throwing motion — basketball one-hand jump shot.

these illustrations may be compared visually, only a force and moment analysis and a comparison of the relative angular velocities and accelerations of each body segment can clearly point out the differences in the throws. The timing of the sequence of deceleration, the range of the segment motion, and the muscle requirements at each joint are needed to fully compare all the motions presented.

Strength of Position

The use of a body segment may be restricted to maintain better ball control, or to make an adjustment that will reduce the muscle needs at a particular joint. The baseball type throw of a soccer goalie, for example, is usually changed to a straight arm throw when long distance is desired. This is because the ball weight is almost three times greater than a baseball and produces very large moments of force at the elbow joint. The weight of a javelin (800 gm) is almost twice that of a soccer ball, and the forces produced by the weight of the javelin and by the fast motion make the elbow very susceptible to injury. Therefore, the elbow joint is generally not bent as much as it could be if the weight were lighter (Fig. 8-14).

The forms adopted in throwing the discus, shot, and hammer are also determined by the relative weights of the objects thrown and by the strengths of the muscles and positions of the segments involved. Other factors may affect the throwing pattern, such as the limited time in football. A good research project would be the comparison of the moments of force, especially at the elbow joint, of the javelin throw, soccer throw, football pass, tennis serve, and baseball pitch (fast ball).

Throwing and Kicking

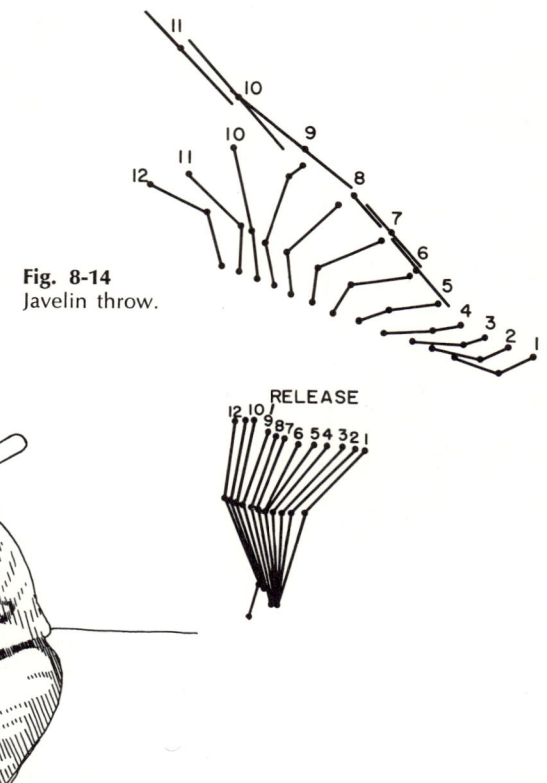

Fig. 8-14 Javelin throw.

Fig. 8-15 Sidearm pitcher with the upper arm in line with a line drawn through both glenohumeral joints.

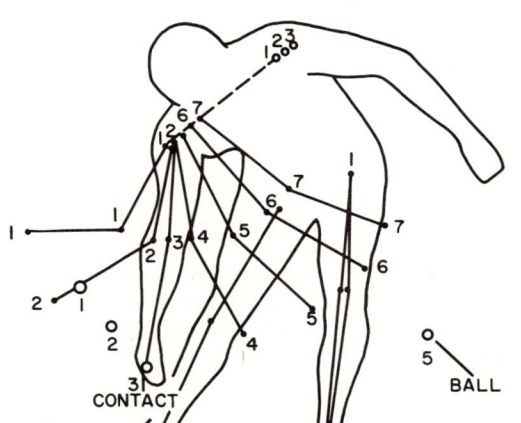

Fig. 8-16 Handball, underarm (ball hit low off back wall) — shoulder and upper arm alignment.

Shoulder Line To maintain a strong throwing position, the upper arm should be either in line with or lower than the line of the shoulders. Figure 8-15 shows the sidearm delivery of a pitcher where the upper arm falls directly on his shoulder line. Figure 8-16 shows the low shot of a handball player with the line connecting the glenohumeral joints tilted downward and the arm below his shoulder line. Figure 8-17a and b shows the upward tilt of the shoulders and the alignment of the upper arm during the javelin throw.

The shoulder line and upper arm relationship has significance for the motion pattern as well as for position strength. The more overhand a pitcher throws, the greater will be: the medial rotation of the upper arm at the glenohumeral joint; the impingement of the vertebrae of the lower back on the discs; and the flexion-extension action of the trunk. In contrast, the lower, sidearm throw of a pitcher places greater emphasis on total trunk rotation. If the moments of force are large at any one joint, the chances for injury or muscle soreness at that point are great. The low back problems due to an exaggerated trunk twist of a tennis serve, the shoulder muscle pulls of a pitcher, and the tendonitis problems at the elbow joint of a pitcher, javelin thrower, or tennis player are well known. A moment analysis could predict future problems, due to a particular style of throwing, or provide a means of comparing form during a slump with past good performances.

Fig. 8-17a, b
Javelin throw — rear view showing upper arm and shoulder line relationship.

Throwing and Kicking

KICKING

When analyzing kicking motions one must consider the pattern of the motion and the foot position at impact. Both will vary according to the position of the ball relative to the nonkicking foot, available time, angle of approach, and the part of the foot being used for contact.

Ball Velocity

Hubert Vogelsinger, the Yale soccer coach, performed specifically for the author, and was filmed using the instep kick on a soccer ball and a football using both the side and straight approaches, as well as the football toe kick (Figs. 8-18 through 8-22). Vogelsinger (I-82) has played both sports profession-

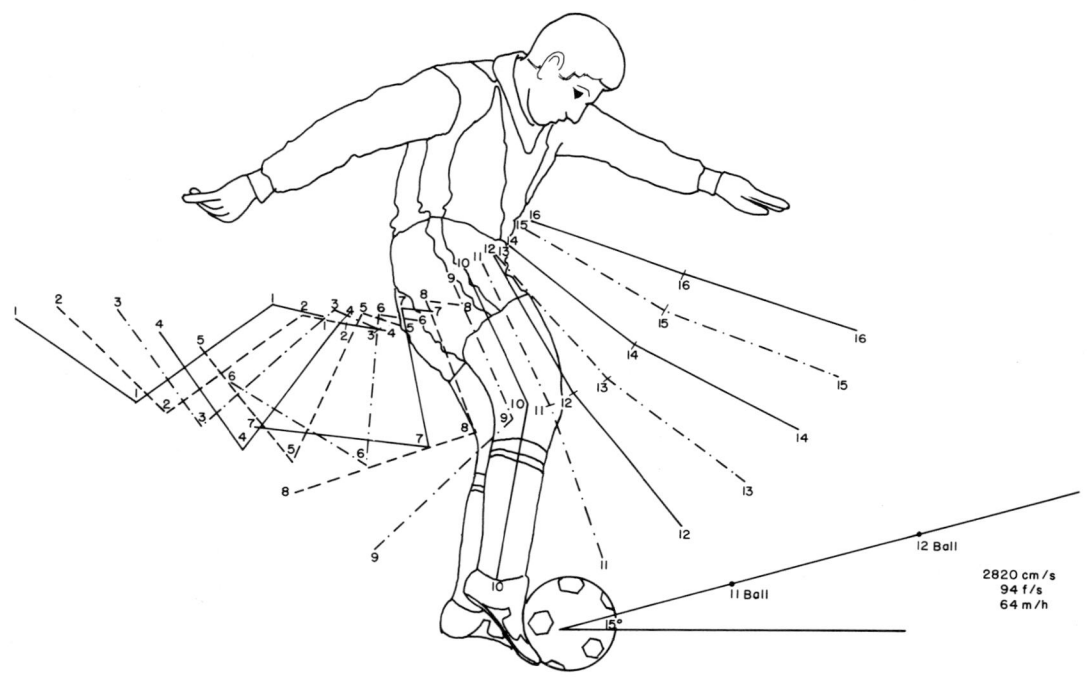

Fig. 8-18
Vogelsinger soccer side approach (66 f/s).

ally, and use of this one kicker eliminates variations in body build and kicking techniques, thus enabling one to better compare force and moment analyses. Several of his kicks using each style were filmed to obtain an average ball velocity. The ranked results in Table 8-1 show that the side approach produced a greater ball velocity than did the straight approach. There are two factors involved in this: (1) foot velocity at impact, and (2) firmness of the foot at impact.

Throwing
and
Kicking

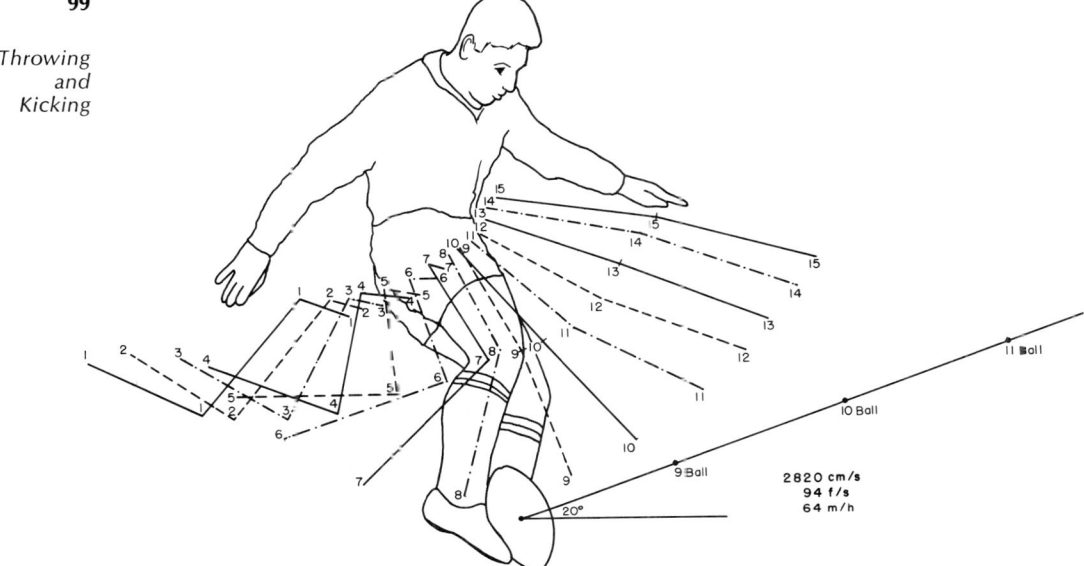

Fig. 8-19
Vogelsinger football side approach (66 f/s).

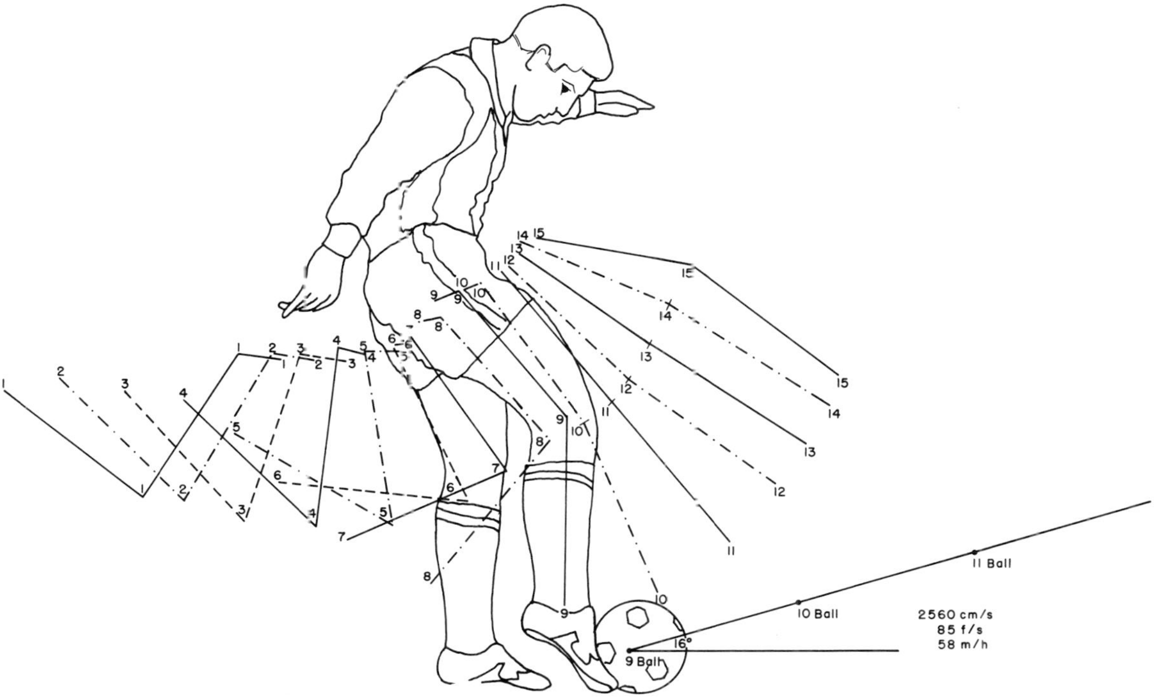

Fig. 8-20
Vogelsinger soccer straight approach (66 f/s).

100

Throwing and Kicking

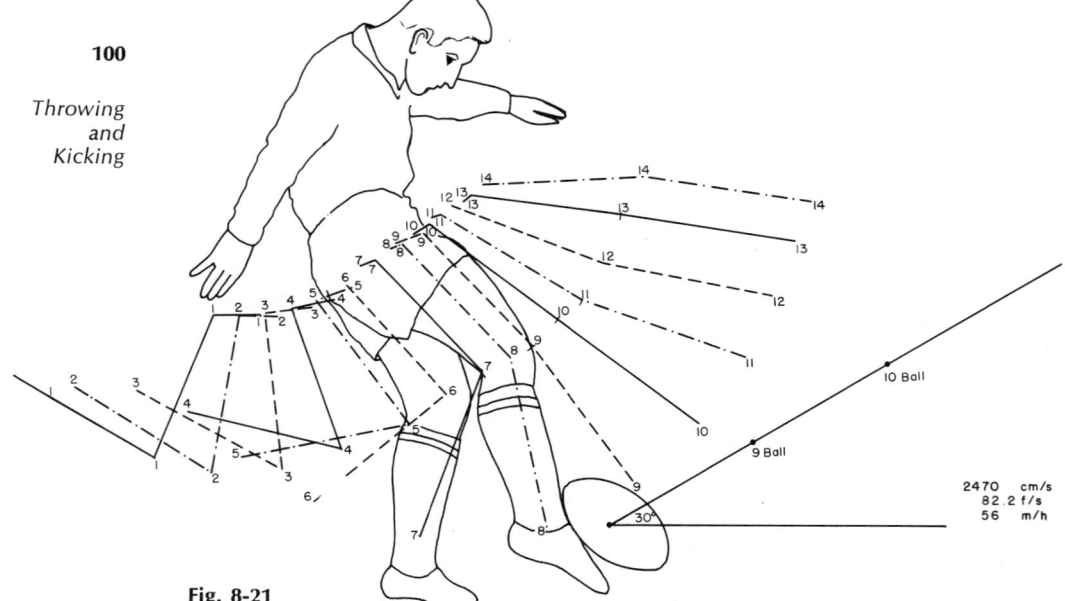

Fig. 8-21
Vogelsinger football straight approach (66 f/s).

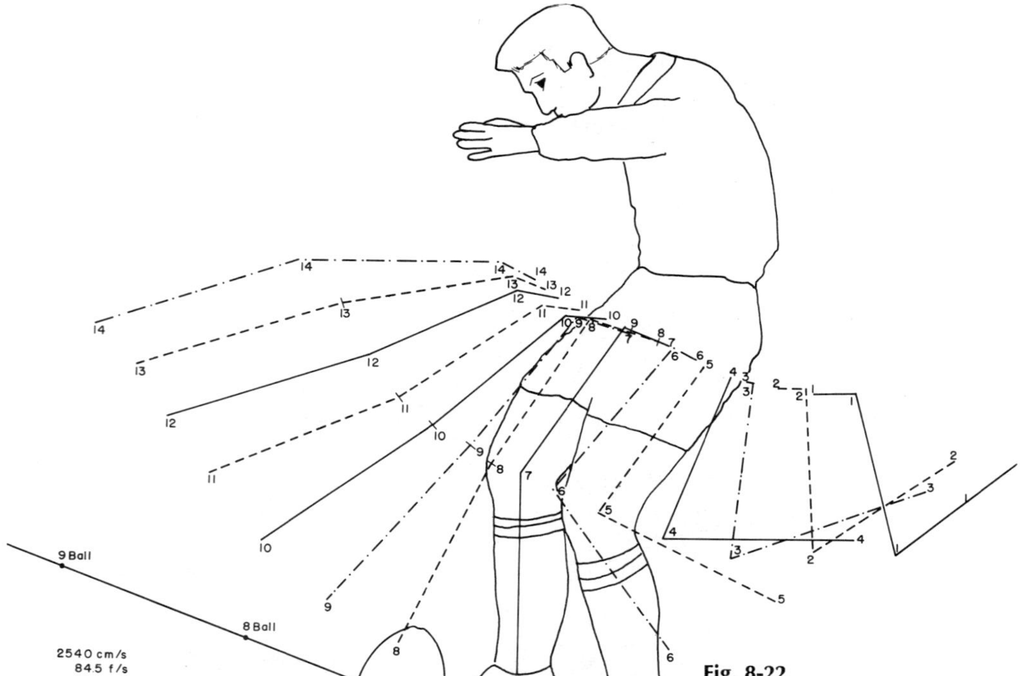

Fig. 8-22
Vogelsinger football toe kick (66 f/s).

Table 8-1 Kicking—Ball Velocities

TYPE OF KICK	RANK	BALL VELOCITY (average of 5 kicks)
Football instep — side approach	1	95.5 ft/sec
Soccer ball rolling — side approach	1	95.5
Soccer ball stationary — side approach	3	91.7
Soccer ball cross	4	88.0
Soccer ball rolling — straight approach	5	87.6
Football toe kick — straight approach	6	82.0
Football instep — straight approach	6	82.0
Soccer ball stationary — straight approach	8	78.2

Foot Velocities Foot velocities just before impact were measured from the motion pictures; the ranked averages of five kicks of each style are presented in Table 8-2. The results indicate that the greatest foot velocity did not always result in the greatest ball velocity, as evidenced by the toe kick and straight approach kicks. Therefore, disregarding how the foot velocity was produced, it can be concluded that the side approach is no better than the straight approach for obtaining maximum foot velocity.

Table 8-2 Kicking—Foot Velocities Before Impact

TYPE OF KICK	RANK	FOOT VELOCITIES BEFORE IMPACT (average of 5 kicks)
Football instep — side approach	1	79.2 ft/sec
Football toe kick — straight approach	2	78.1
Soccer ball rolling — straight approach	3	70.7
Football instep — straight approach	4	66.7
Soccer ball rolling — side approach	5	65.0
Soccer ball stationary — straight approach	5	65.0
Soccer ball stationary — side approach	7	64.2
Soccer ball cross	8	53.6

Foot Striking Mass To obtain data pertaining to the striking mass, the foot velocity was also measured after impact; this made it possible to use the conservation of momentum formula. The ranked results are presented in Table 8-3. The wide range for each kick shows how difficult it is to contact the ball firmly and consistently. Vogelsinger more consistently hit a football with a straight approach than with either the side approach or toe kick; he was more consistent with the soccer side approach than the straight approach. Although these data would differ for another kicker and with a larger number of kicks measured, the important conclusion is that *the placement of the foot on the ball is a greater variable than is the attainment of maximum foot velocity.*

Table 8-3   Kicking—Striking Mass

TYPE OF KICK	RANK	AVERAGE STRIKING MASS	RANGE (lbs)
Soccer ball stationary — side approach	1	8.6 lb	5.7 – 10.6
Football instep — straight approach	1	8.6	8.5 – 8.7
Soccer ball cross	3	8.3	3.5 – 13.0
Soccer ball rolling — side approach	4	7.0	4.0 – 11.0
Soccer ball stationary — straight approach	4	7.0	3.3 – 12.6
Soccer ball rolling — straight approach	6	5.7	3.6 – 9.6
Football instep — side approach	6	5.7	3.2 – 8.0
Football toe kick — straight approach	8	5.3	4.2 – 7.0

Moments of Force in Kicking

One kick of each style was then analyzed to determine the joint muscle action. The ball velocity and angle of each kick are given in Table 8-4.

Table 8-4   Kicks Analyzed

TYPE OF KICK	ANGLE OF KICK (in degrees)	BALL VELOCITY (mph)
Soccer side approach	15	64
Football side approach	20	64
Soccer straight approach	30	56
Football straight approach	16	58
Toe kick	21	58

The joint moments of force, due to the acceleration of trunk rotation, were nearly equal for all kicks except the toe kick; the toe kick had the largest moments due to trunk rotation deceleration. It is significant that the deceleration of trunk rotation was greatest for all three straight approach kicks. Thigh flexion of the kicking leg was greatest for the straight approach kicks, whereas thigh deceleration patterns were not consistent when the side and straight approaches were compared. It is also significant that the two kicks that had the lowest thigh deceleration showed the greatest knee extension. This points out the influence of one segment on another and the fact that if more muscle force is used at one joint, less will be required at the adjacent joint to obtain equal foot velocity. It also indicates that if the moments at one joint are small, the muscle forces at another joint will be greater than necessary. A thorough understanding of the influence of one segment upon another explains not only the muscle needs for a given performance, but the reason for possible injury.

Characteristics of Good Kicking

The proper timing of the accelerations and decelerations of each segment (with a range of motion sufficient to allow this timing) and a solid impact of the foot will produce maximum ball velocity. Thus the characteristics of a good kick are:

1. sufficient trunk rotation to allow a fully "back" position of the kicking thigh, and
2. a fully flexed knee joint to allow a complete swing of the lower leg.

It is also important that the nonkicking foot be placed relative to the ball so as to allow the best fit of the kicking foot to the ball. The shape of the soccer ball makes it easier to contact more firmly using the side approach than the straight approach. The toe will meet the ball first if the ball is too far forward of the nonkicking foot, thus deadening the impact and probably increasing the altitude of ball flight.

Ball Position

Kicking may be further classified by the ball's position either on or off the ground. Punting is the only off-the-ground kick in football, while in soccer many variations exist for off-the-ground balls. A solid foot position on the ball is easier to attain in soccer when the ball is off the ground because the foot will fit the ball without the necessity of being fully extended. Many players can kick well if the ball is only an inch or two off the ground, but cannot kick a stationary or rolling ball well. After contact, the flight of the ball must be low if shooting, or high if clearing. Therefore, adjustments in the total body position and foot positions are necessary for accuracy of the flight.

Toe kicks are identical in football and soccer, as is the actual kicking motion when using the instep to contact the ball. Kicking differences occur because the ball is generally moving at various angles relative to the kicker in soccer, or the kicker may be trying to put spin on the ball.

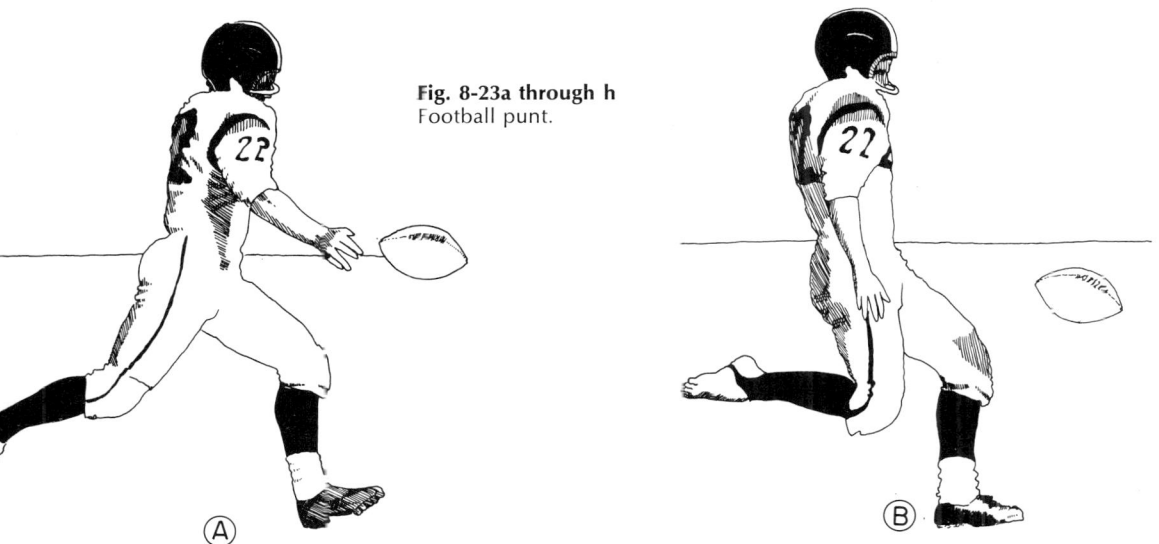

Fig. 8-23a through h
Football punt.

Fig. 8-23 continued

Teaching of kicking skills at the secondary school level can only be accomplished when more teachers understand the mechanics of kicking and the importance of the foot's placement on the ball. It is nearly impossible to see the actual foot placement during a kick, due to the speed of movement of the foot. The slower-moving beginning and ending of the leg motion produce a visual image of what happens at contact, but this is often misleading. The only way to see actual foot position is to take slow-motion pictures of each player.

Figures 8-23 through 8-37 are presented to clarify the good and poor aspects of kicking, and to especially aid soccer coaches in the teaching of kicking. These illustrations are discussed in more detail in the succeeding pages.

ILLUSTRATED KICKING ANALYSIS

Punt (Fig. 8-23a–h)

Figure 8-23a through h shows a punter who consistently kicked ten yards further without his shoe. Inasmuch as his motion did not change (no increase in foot velocity without the shoe), the firmness of the extended foot was better with the shoe off. Apparently he could not fully extend his foot with his shoe on, and the force of the impact extended it further for him. This added "give" of the shoed foot reduced the striking mass and thus the yardage. The length of the last stride before kicking controls the amount of hip rotation and the position of the thigh in its full back position (Fig. 8-23a). As the thigh moves forward, the full knee bend generally occurs when the thigh's longitudinal axis is perpendicular to the ground (Fig. 8-23b). Deceleration of the thigh aids the increase in velocity of the lower leg (Fig. 8-23c and d), so maximum foot velocity is attained at impact. Contacting the ball high on the foot produced an end-over-end flight (Fig. 8-23e and f), and the foot reacted to the reduced force after impact by flexing (Fig. 8-23g). The high follow-through allows a gradual deceleration of the motion; thus less muscle is needed for stopping the action, and the foot extends again as a result of stretching the foot extensors due to the straight leg position shown in Fig. 8-23h.

Toe Kick (Soccer Goal Kick) (Fig. 8-24a–e)

The toe kick requires a straight approach. The shoulders remain almost perpendicular to the line of the run throughout the whole movement. Hip rotation depends upon the length of the last stride (Fig. 8-24a), and the full bend of the knee occurs when the thigh is approximately perpendicular to the ground (Fig. 8-24b). The thigh decelerates to aid the swing of the lower leg (Fig. 8-24c), and contact is made with a firm foot perpendicular to the lower leg; the ball is well forward of the nonkicking foot (Fig. 8-24d). The length of

106

Throwing and Kicking

Fig. 8-24a through e
Soccer toe kick.

Throwing and Kicking

Fig. 8-25a through f
Soccer straight approach kick.

Fig. 8-26a, b
Soccer straight approach — foot placement.

the follow-through is governed by the position of the ball at contact relative to the nonkicking foot. The further forward the ball, the shorter the follow-through; this is because the total body center of gravity is forward of the support leg, thus requiring the kicking leg to get back on the ground to maintain body balance (Fig. 8-24e).

Soccer Straight Approach (Fullback Goal Kick) (Fig. 8-25a–f; Fig. 8-26a and b)

This instep kick was chosen because of its similarity to the toe kick of Fig. 8-24: the straight approach with the "square" shoulders, full back position of the thigh due to hip rotation (Fig. 8-25a), perpendicular thigh at full knee bend (Fig. 8-25b), pointing of the thigh at the ball during the fast swing of the lower leg (Fig. 8-25c), position of the ball forward of the nonkicking foot (Fig. 8-25d), and the firm foot at impact. The follow-through (Fig. 8-25d–f) is higher than in the toe kick due to the body adjustment, and the knee is bent at the finish to reduce the stretch on the muscles of the back of the leg. Figure 8-26a shows the rear view of the foot at contact, and Fig. 8-26b shows the top view at impact.

Throwing and Kicking

Fig. 8-27a through g
Soccer angled approach kick.

Soccer Angled Approach (Fullback Goal Kick) (Fig. 8-27a–g; Fig. 8-28)

An approach angle of about 30° to 40° to the intended ball flight requires that the nonkicking foot be turned to point in the direction of the intended flight. The whole trunk and kicking leg revolve about the left hip, and the approach angle makes it easier to obtain a fuller back position of the right thigh than when using a straight approach (Fig. 8-27a). The trunk and thigh rotate together, almost as one segment, until the full knee bend is reached (Fig. 8-27b). The line of the shoulders never becomes perpendicular to the ball flight direction, so the remainder of the kicking motion is very similar to the straight approach (Fig. 8-27c–g). The rotation of the hips relative to the shoulders results in a slightly twisted body position at impact with the foot angled slightly across the ball (Fig. 8-28).

Fig. 8-28
Soccer angled approach — foot placement.

Soccer Cross (Figs. 8-29a–c, 8-30, 8-31a–f, and 8-32a–f)

The approach angle of the cross (ball flight approximately 90° to the approach direction of the run) makes it easier to get a full back position of the kicking leg, but much more difficult to obtain a large foot velocity in the direction of the ball flight. It is also difficult to maintain good body balance and get the nonkicking foot turned sufficiently to aid the turning of the body and kicking of the leg during the swing phase. The foot position at contact (Fig. 8-30) is across the ball, because the ball continues to roll away from the nonkicking foot after it is placed. Figure 8-29a–c shows a cross made off a fast run where it is difficult to turn the nonkicking foot fully, and Fig. 8-31a–f shows the same kicker in a top view. Figure 8-32a–f presents a perfectly turned nonkicking foot, excellent body balance, and the full kicking action of a well-executed cross made from a slightly slower approach run.

Throwing and Kicking

Fig. 8-29a, b, c
Soccer cross.

Fig. 8-30
Soccer cross — foot placement.

Fig. 8-31a through f
Soccer cross — top view.

112

Throwing and Kicking

Fig. 8-32a through f
Soccer cross — non-kicking foot fully turned.

Fig. 8-33a through e
Faulty soccer kicking style.

Fig. 8-34
Poor foot placement for instep kick (rear view).

Soccer Angled Approach (Faulty Style) (Figs. 8-33a–e and 8-34)

Figure 8-33a–e is shown to illustrate the most common kicking techniques used by many high school and college players who have not yet learned to kick properly. The foot is placed on the ball with the toe to the outside and the heel low, so the inside of the instep contacts the ball and the excess knee bend of the nonkicking foot results in a faulty sitting position (Fig. 8-34). This technique usually results in shots missed over the goal or crosses kicked behind the goal.

Foot Placements at Impact in Football (Fig. 8-35a and b, 8-36a–d, and 8-37a and b)

The kicking motions are very similar in soccer and football, but the foot placement on the ball varies. Successful kickers use the toe, instep, and inside of the foot in football. Figure 8-35a and Fig. 8-35b show the instep kick and Fig. 8-36a–d, the inside of the foot kick. The inside of the foot kick sometimes produces an unstable flight path if the ball is contacted too high (almost directly behind the ball center of gravity,) as no end-over-end spin is produced. Figures 8-37a and 8-37b show the varying positions of the ball relative to the nonkicking foot for the toe kick.

Fig. 8-35a, b
Football kick —instep contact.

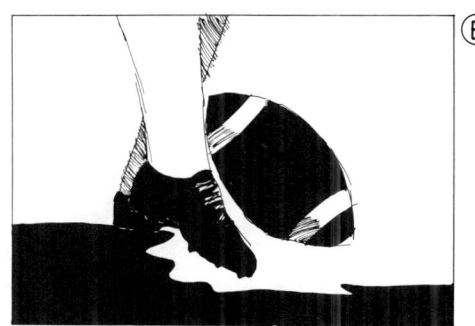

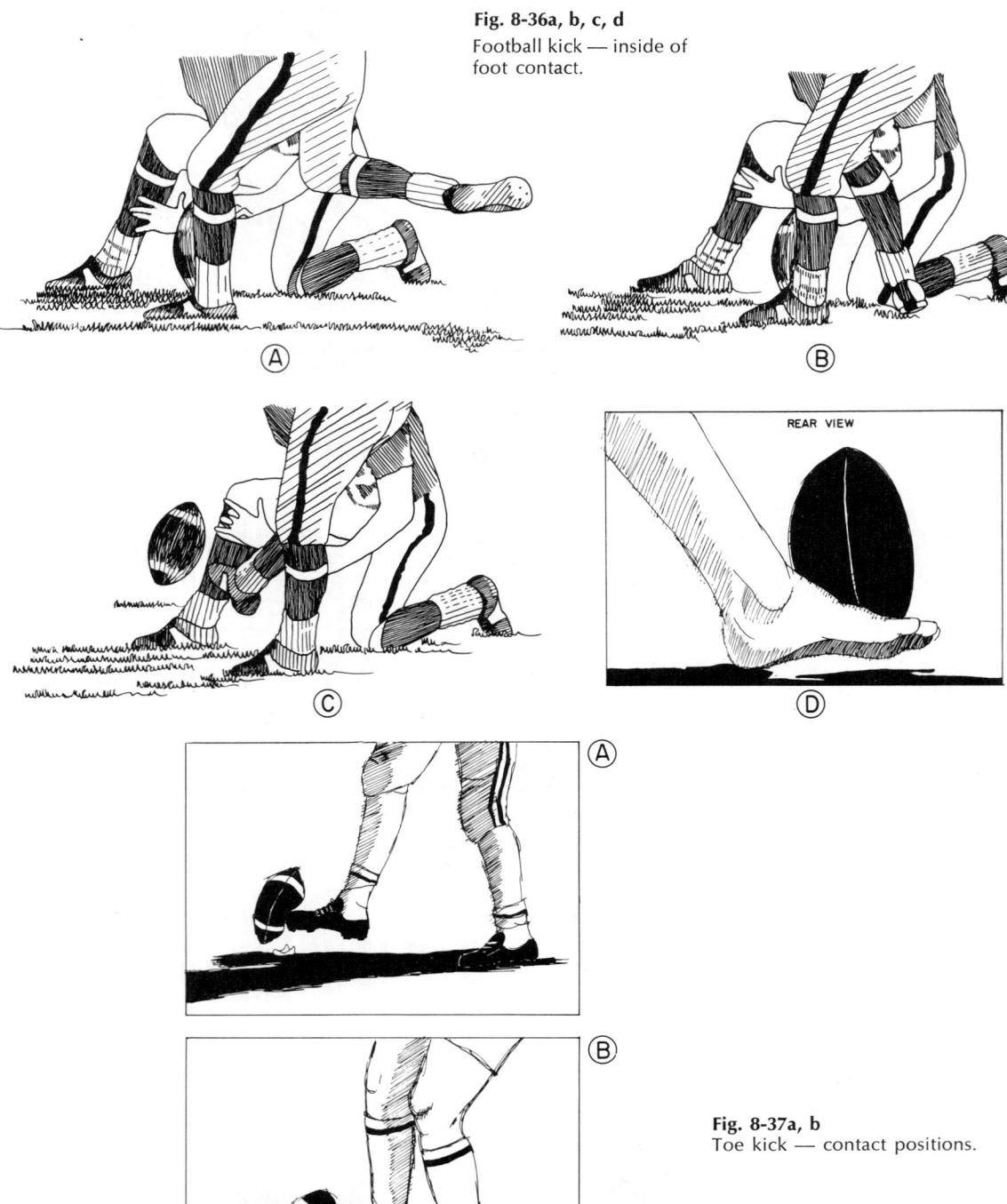

Fig. 8-36a, b, c, d
Football kick — inside of foot contact.

Fig. 8-37a, b
Toe kick — contact positions.

chapter 9

ANALYSIS OF SELECTED SPORTS

This chapter discusses a few selected sports involving problems of analysis that are representative of the many problems encountered when a wider variety of body motions are studied. Of course, every motion analyzed has its own special problem — it may be more important to know the body segment position than the total body center of gravity, or knowing the path of a joint center or the sequence of timing of the segments may be more helpful than knowing the muscle involvement. However, in any force and moment analysis the link system must be determined, the external forces measured, and the forces of those body parts not in the chosen link system calculated.

The computer program in Appendix B is written to handle seven body segments linked from the fixed end to the last moving segment. A choice must be made as to which body segments will constitute the link system during a three-dimensional motion that is nonsymmetrical (as in a throwing motion). How does one analyze the action of both arms and legs during a take-off of the broad jump? How do you analyze the badminton or tennis smash if the body motion takes place while the player is jumping backward from the right to the left foot? (Or jumping forward from the right to left foot during the lacrosse shot or the running football pass?) How can one analyze the volleyball spike or swimming stroke where there is no fixed point? What is the link system during the pole vault or golf swing when both arms are attached but performing differently? How is the motion analyzed when sitting with both the hands and feet touching outside objects (a closed chain, as in rowing and kayaking)? The following sports analyses should help answer these questions.

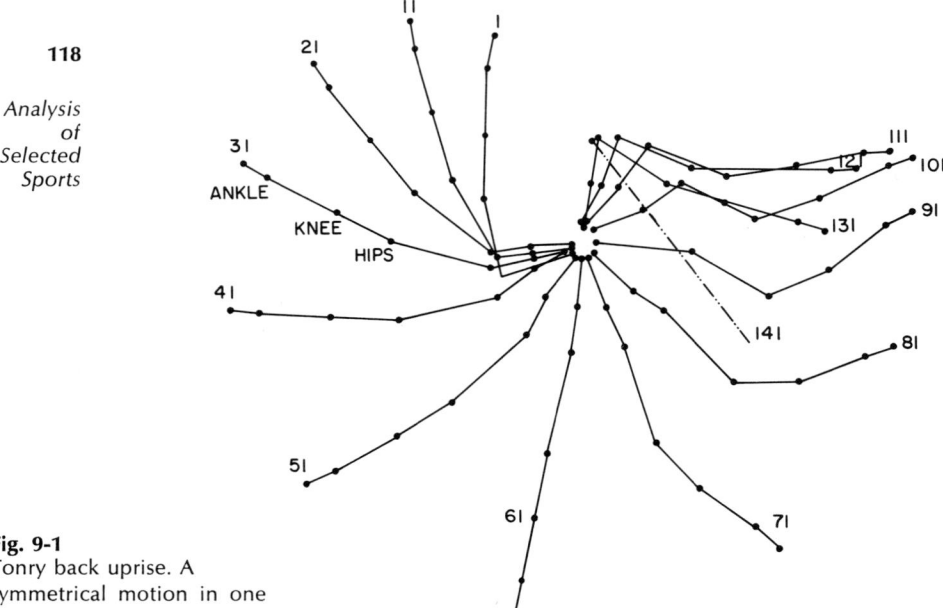

Fig. 9-1
Tonry back uprise. A symmetrical motion in one plane (64 f/s).

GYMNASTICS

Gymnastics has some of the easiest and some of the most complex motions to analyze. There are many symmetrical motions with either the hands or feet remaining fixed. The back uprise on the horizontal bar (Fig. 9-1), the head spring (Fig. 9-2), and the "peach" to a handstand (Fig. 9-3a–d) are representative of this group. The peach to a handstand produces a tracing with many overlapping lines if the tracing is done on one sheet of paper, so the motion is broken into four parts. Even though a hand release occurs, the desired information is easily obtained. The path of the total body center of gravity is also drawn so that a comparison may be made with the performance of another gymnast.

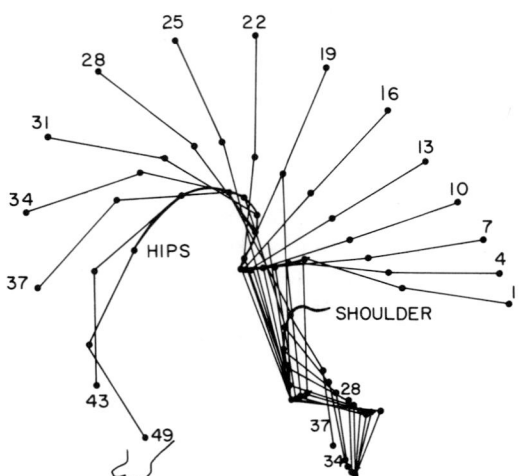

Fig. 9-2
Head spring. A symmetrical motion with the elbows moving out of the place of motion (64 f/s).

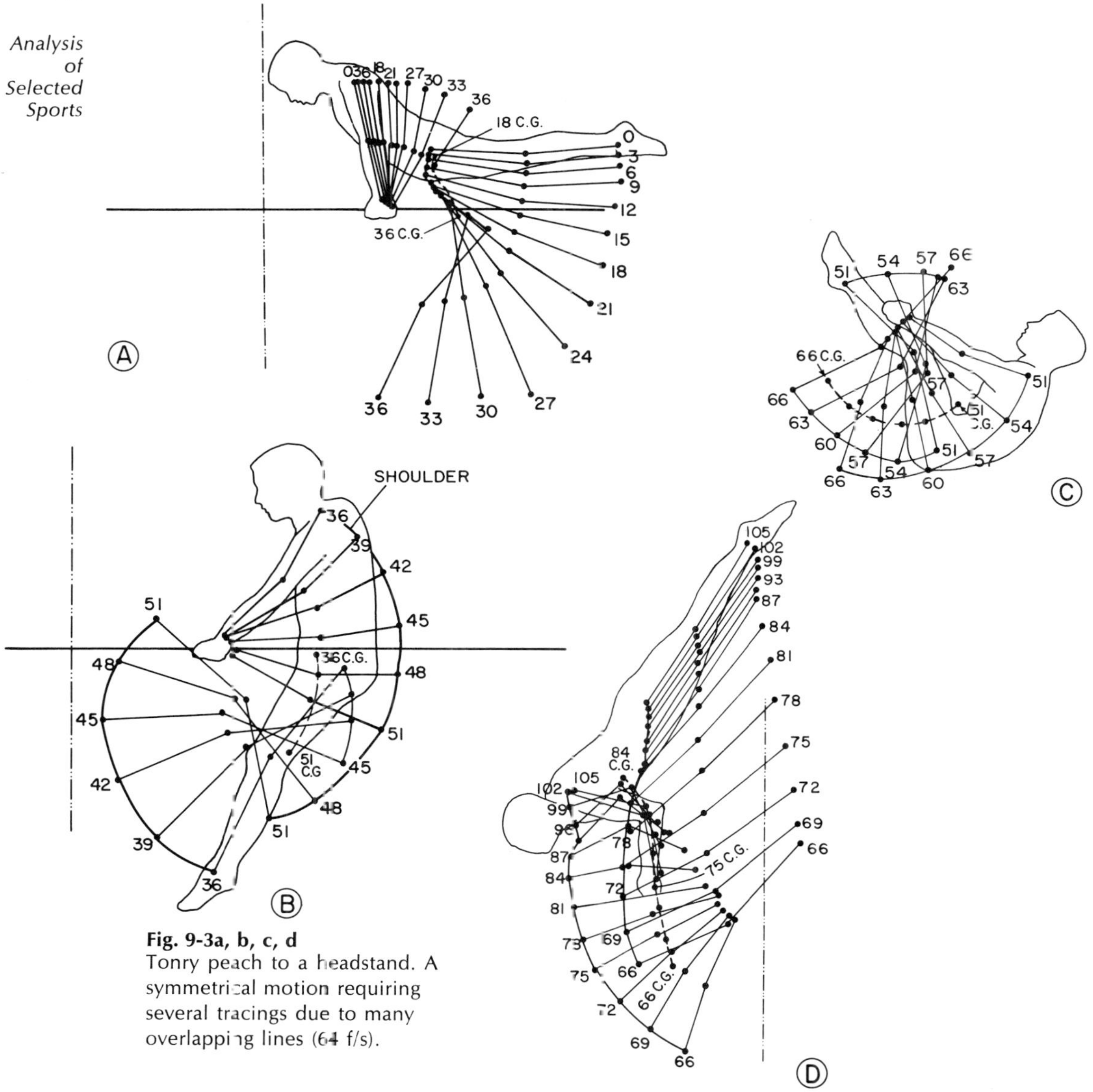

Analysis of Selected Sports

Fig. 9-3a, b, c, d
Tonry peach to a headstand. A symmetrical motion requiring several tracings due to many overlapping lines (64 f/s).

Knowing the direction and magnitude of the force on the hands may be more important than knowing the extent of the muscles used. Dusenbury (I-20) analyzed the front and reverse giant swing on the still rings and found that the reverse giant produced 35% less maximum vertical force on the hands. Therefore, the reverse giant should be taught first to lessen the chance of falling off, if this is a factor in an individual's progress. Dusenbury also concluded that the front giant was easier to perform if only strength requirements are considered. Proper timing of the body segments produces lower joint moments, so two performers can be compared for efficiency of motion on this element.

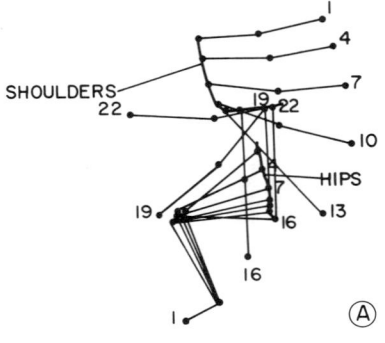

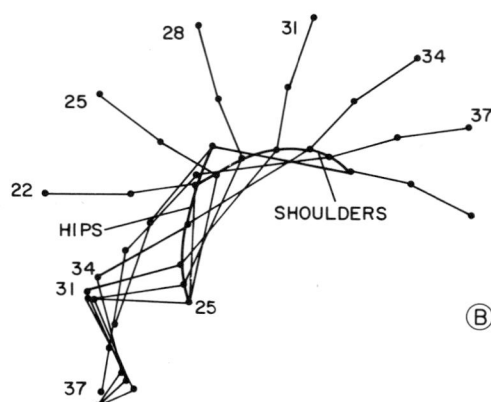

Fig. 9-4a, b
Tonry standing back handspring.

Figures 9-4a and b and 9-5 show a standing back handspring and a back handspring out of a roundoff. The paths of the joint centers show the differences between the two motions, and a force and moment analysis indicates the extent to which the arms aid the take-off. A comparison could also be made between a back handspring done during competition in early and late season to determine any major changes which improved performance.

Mid-air motions present a problem when analyzing muscle usage, especially when the movement is a twist off a somersault (such as in a "late fliffis" done on a trampoline). A mid-air joint moment analysis can be done only by determining the joint which is moving the least and analyzing other body motions relative to this most stable joint (usually the hip joint, as shown during the soccer head). If all body parts are moving a great deal, only the axis of rotation, the path of the total body center of gravity, the path of the joint centers, or body segment positions relative to the floor may be determined (such motions

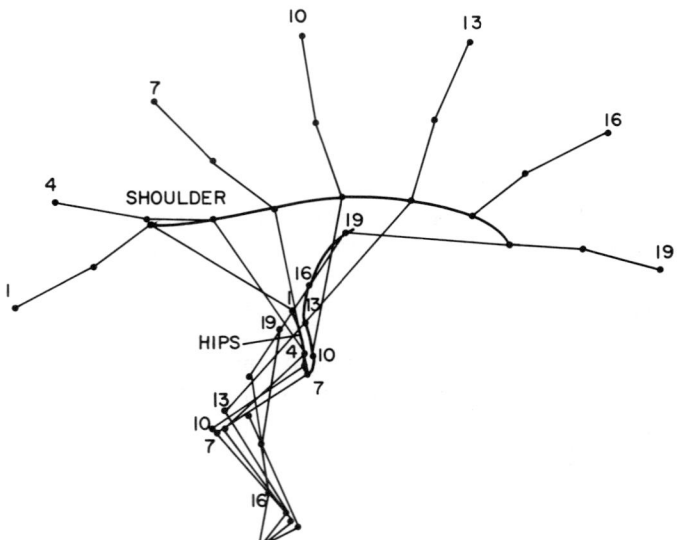

Fig. 9-5
Tonry back handspring out of a roundoff. The tracings in Figures 9-4a, b and 9-5 are used to compare the differences between the two techniques.

occur in tumbling, trampolining, apparatus mounts and dismounts, and vaulting).

The motion just before take-off may be very important, as the timing of the maximum upward force ($+F_y$) must be coordinated with the take-off. On a trampoline, the upward force must be coordinated with the restitution of the bed. (This is also true in diving and pole vaulting, where the upward force coordinates with the straightening of the board and pole, respectively.)

There is a need in gymnastics for the study of the body motion forces which cause injuries. It is easy to get slow motion pictures of a good performance, but an injury is recorded for analysis only by chance. The improper motions producing the high forces causing falls, as well as muscle and joint injuries, need further study.

SWIMMING

Swimming presents two special problems for analysis: there is no fixed point, and the external forces due to movement through water are very difficult to measure. Fortunately, a great deal can be learned from patterns of the motions with the forces disregarded. Because there is no stationary point, the arm analysis is done from the hips up, and the leg analysis from the shoulders down.

The crawl tracings of Schollander (Fig. 9-6a–c), Mosconi (Fig. 9-7a–d), and a national intercollegiate 50 yd. champion, James Van Kennen (Fig. 9-8a–c) are presented for comparison. The angles of the body segments, the paths of the joint centers, the time of the hand in the water, and the relationship of the hand movement to the body and to the water may all be determined. The tracings are of the actual motion as the swimmer passes a stationary camera (but Fig. 9-7d is traced by moving the paper for each frame to place the hip joint in the same spot — this obtains the pattern of motion of the arms and trunk relative to a fixed point). The tracing in Fig. 9-7d shows that the arm is moving backward relative to the hips during the whole stroke except for the first four positions after entry. Figures 9-6a, 9-7a, and 9-8a show arm movement relative to the water with the arms moving forward after entry, causing drag, until the forearm angle reaches approximately 30°. The movement of the elbow and wrist joints is continuously backward after that position until exit.

The front views show how the elbow is partially bent during the motion, but the upper arm angle remains almost constant. This bending of the elbow reduces the muscle force needed for arm extension, but at the same time increases the force needed for medial rotation of the upper arm. The changing path of the hand and forearm keeps as much surface area near the optimum vertical position as long as possible. The arm is then pulled out of the water without causing any drag, thus constantly changing the hand angle so it remains vertical. The drag during the beginning of the stroke is minimized by moving through the area of forward motion rapidly, as done by Van Kennen (Fig. 9-8a).

The extent of knee bend, thigh angular motion, and time for one kick cycle may also be compared. Schollander's thigh motion (deceleration) aids the lower leg motion, and the maximum shank angle causing drag is only 20° when the foot is up. Van Kennen uses very little thigh motion, but the maximum shank angle is 32°. This decreases thigh drag, increases shank drag, and increases the knee moments of force. Mosconi's kick is very similar to Schollander's except that his downward maximum shank angle is 20° compared to 13° for Schollander.

Analysis of Selected Sports

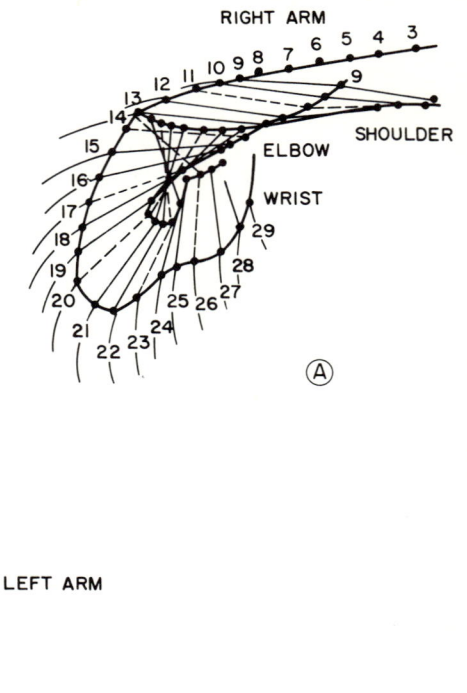

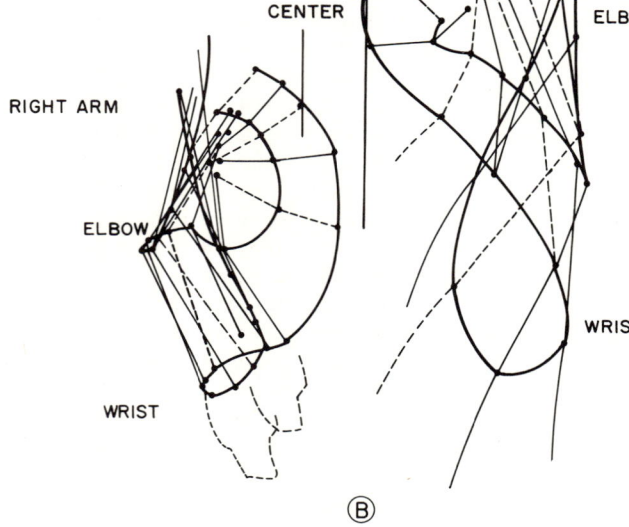

Fig. 9-6a, b, c
Schollander crawl. All swimming pictures were taken at 24 f/s.

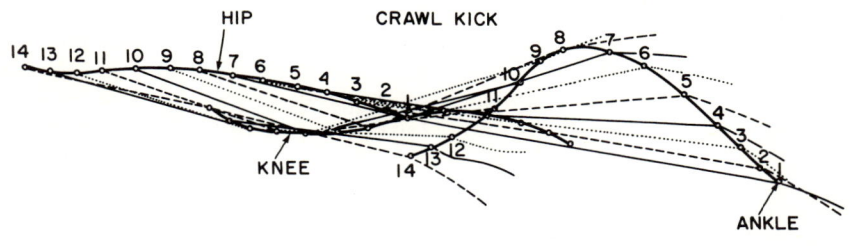

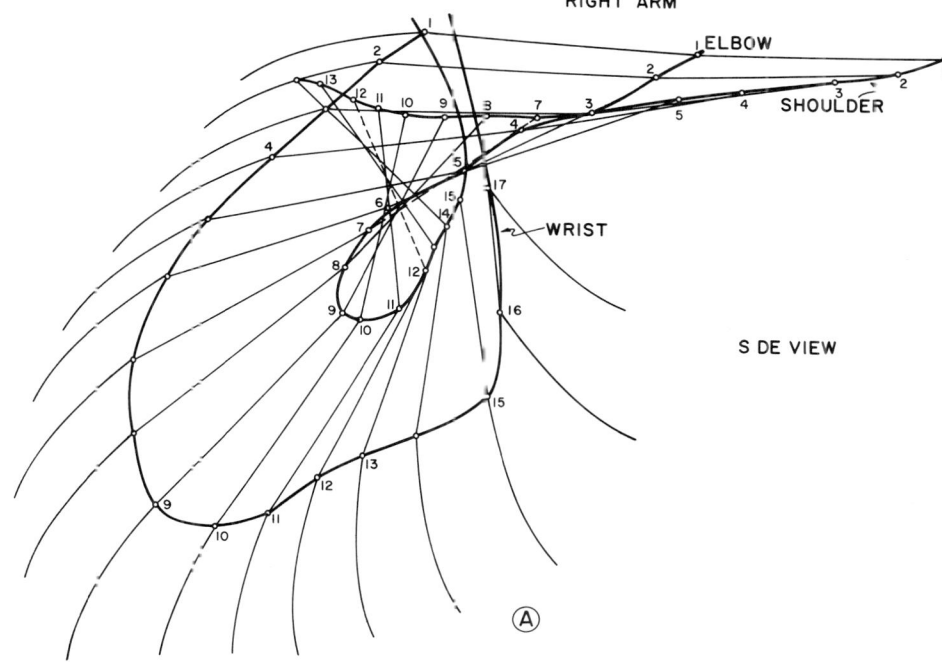

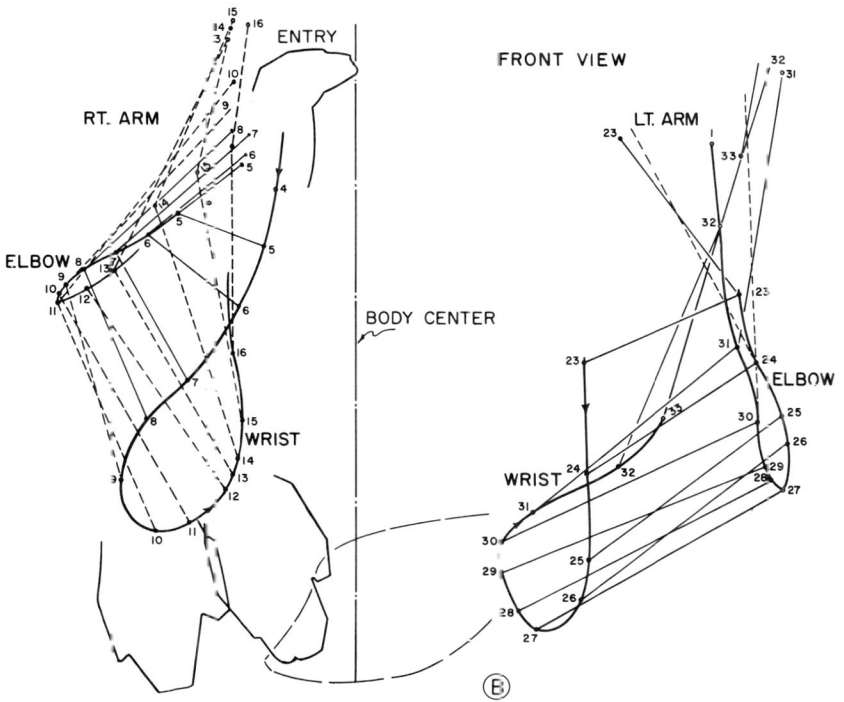

Fig. 9-7a, b
Mosconi crawl.

124

Analysis of Selected Sports

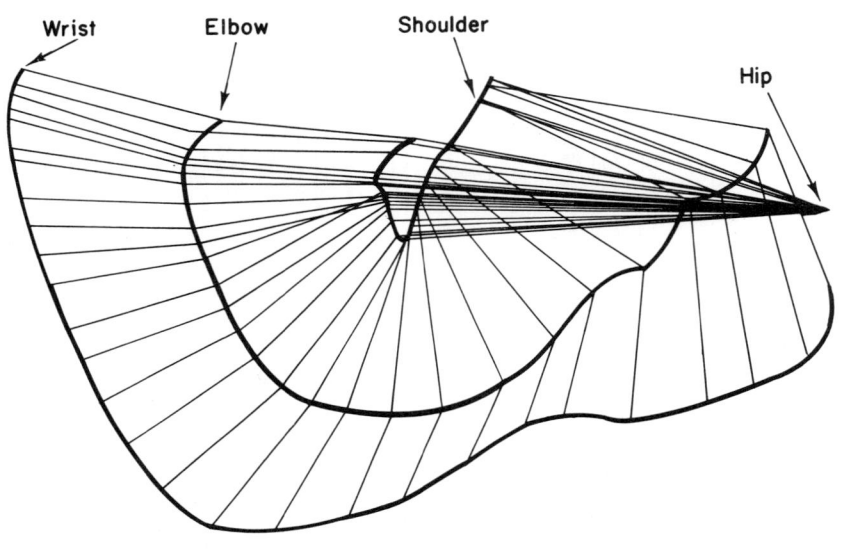

Fig. 9-7c, d
Mosconi crawl.

Ⓓ ARM MOTION RELATIVE TO TRUNK

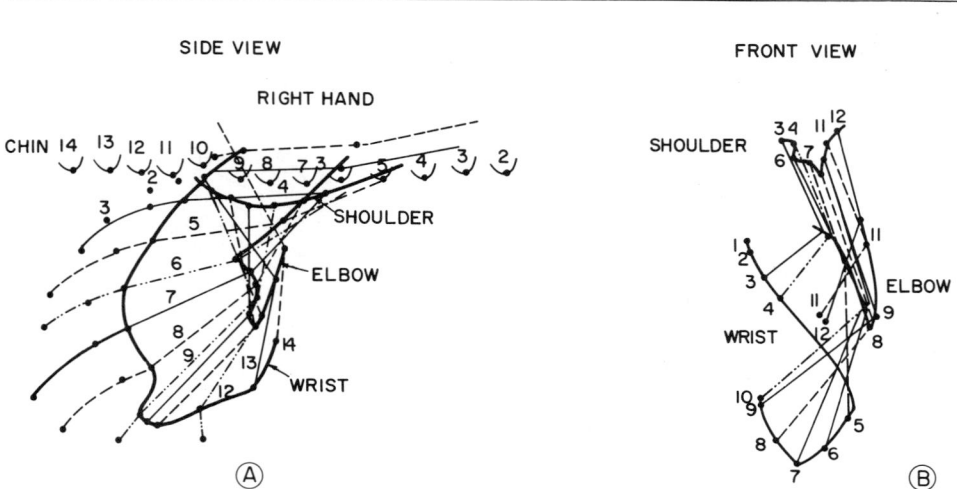

Fig. 9-9a, b, c
Schollander fly.

Fig. 9-8a, b, c
Van Kennen crawl.

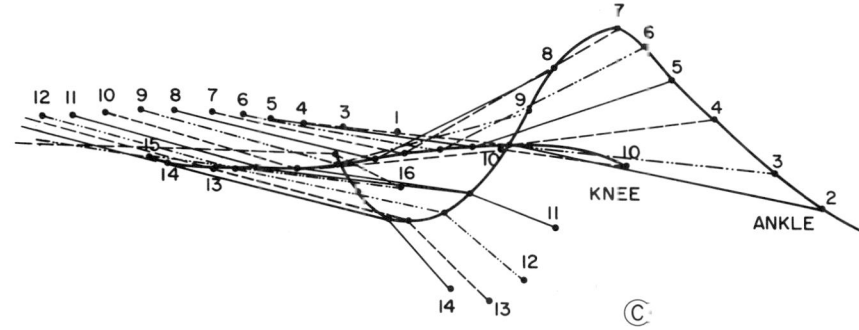

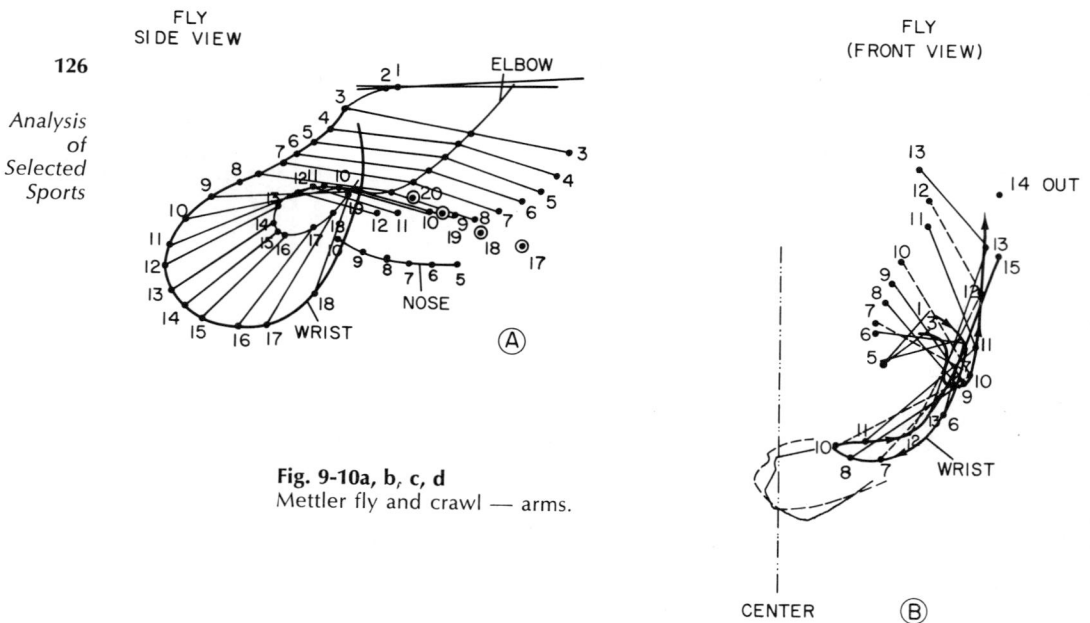

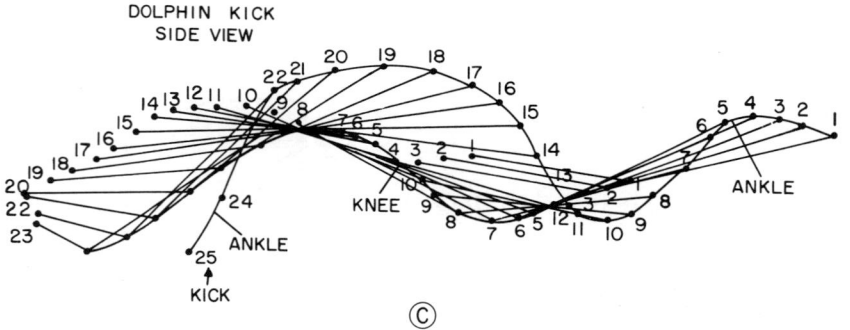

Fig. 9-10a, b, c, d
Mettler fly and crawl — arms.

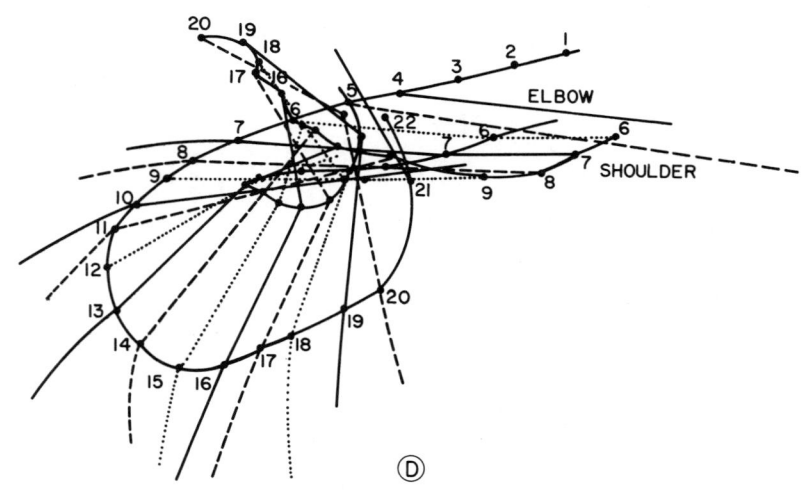

The optimum motion for kicking cannot be determined until drag factors are obtained in various positions for each body segment at various velocities. It cannot be said that any one style of kicking would be best for everyone due to the many individual differences, as well as the coordinated differences, between the arms and legs. If Van Kennen's arm pattern were combined with Schollander's leg pattern, the result would be very close to a composite of Mosconi.

The butterfly strokes are also presented for comparison (Schollander, Fig. 9-9a–c; Mettler, Fig. 9-10a–d; and Mosconi, Fig. 11). (Figure 9-10d presents Mettler's arm motion in the crawl for comparison with Fig. 9-10a.) The wide elbows are characteristic of the "fly," but Mettler's motion has the hands closer to the center line than Schollander's, due to differences in the positions of the forearm (Figs. 9-9b and 9-10b). The patterns of the kicks are very similar, but the smoothness of the path of the hips is lost by Mosconi due to the leg pattern (Fig. 9-11).

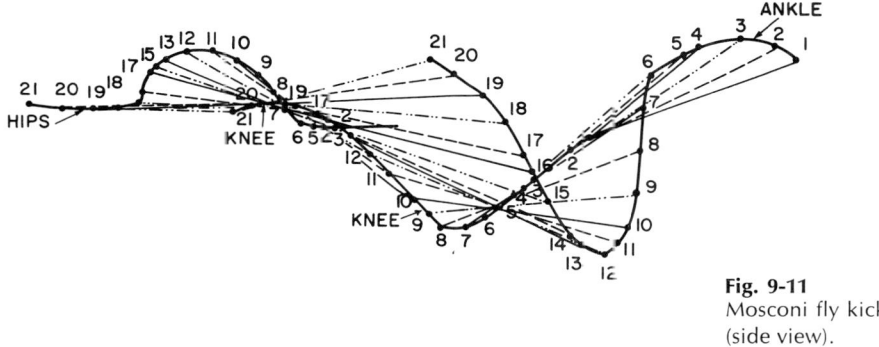

Fig. 9-11
Mosconi fly kick (side view).

The breast stroke tracings of Buckley are presented so that they may be compared with other swimmers (Fig. 9-12a–c). The restricted backward motion of the arms, the coordination of arms and legs, the coordination of the thigh with the lower leg, the body segment positions, and the joint paths may all be obtained from these composites.

When a swimming stroke is analyzed for joint moments of force, the weights of the body parts must be equal to zero due to the water buoyancy. Therefore, the forces due to the pressure of the arms against the water must be measured externally. Alley (I-2) presents data that may be used for a force and moment analysis. He measured the resultant horizontal arm and leg force separately by attaching a measuring device to a belt. However, the changes from one instantaneous arm position to the next are very difficult to measure and a better device is needed to find significant differences in swimming styles. Also, measurements on a stationary swimmer are different from those on a swimmer moving at varying velocities. The moments for swimming were obtained using an average force of 14 lb for the arms and 5 lb for the legs. These data were treated as external forces at the hands and feet in the computer program (Appendix B). This assumes that the hands are the stationary point rather than the hips, and, therefore, the results are approximate until better measurements are available.

128

Analysis of Selected Sports

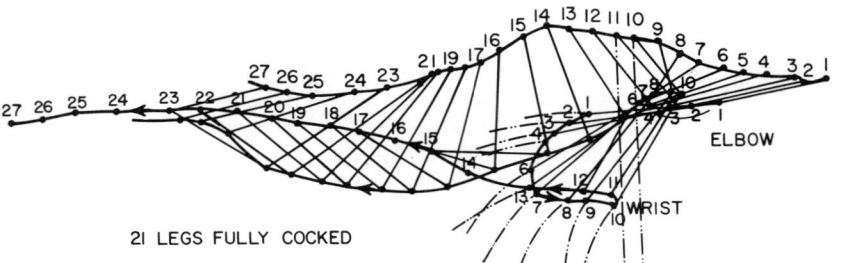

ARMS SIDE VIEW

21 LEGS FULLY COCKED

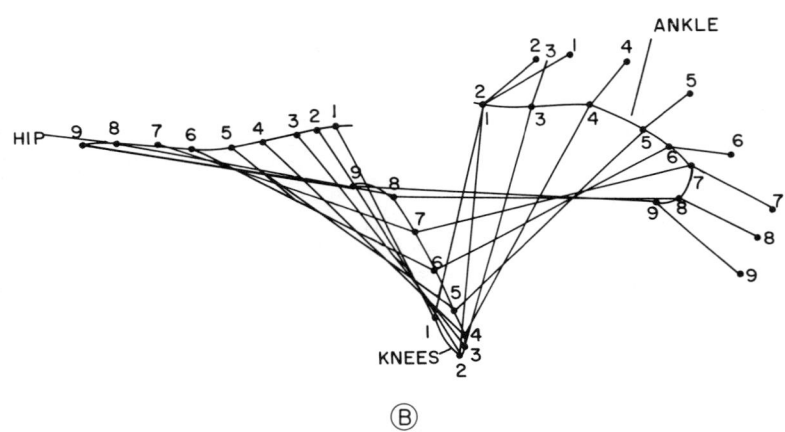

CLOSING KICK — SIDE VIEW

Fig. 9-12a, b, c
Buckley breast stroke.

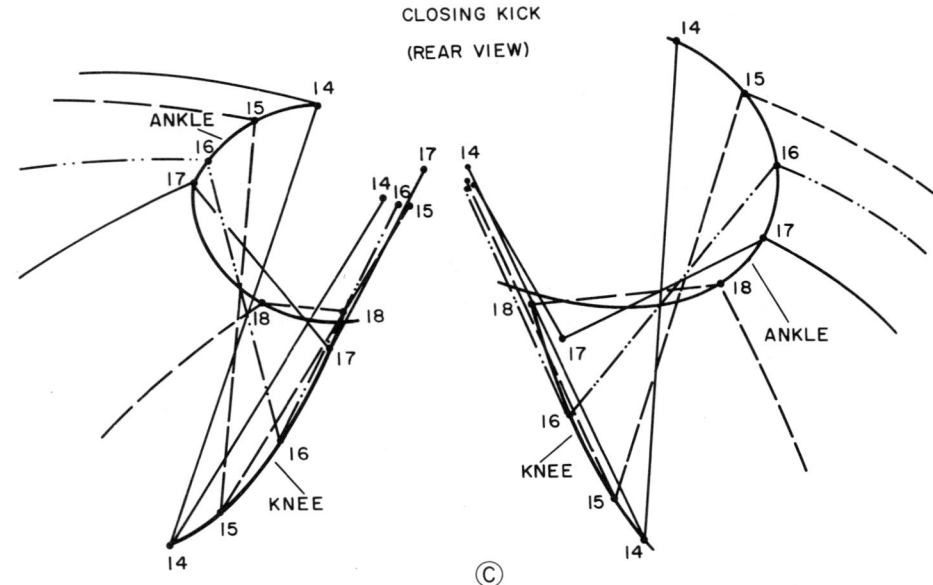

CLOSING KICK
(REAR VIEW)

Kayak paddling is presented because of the many additional problems introduced by its analysis. There are two fixed points, varying external forces at the hands, and a nonsymmetrical motion that must coordinate one side with the other. A paddling style which minimizes the joint moments is desired due to the importance of muscular endurance during racing. Fluid dynamics pertaining to boat zigzag, paddle position, and paddle movement relative to the kayak and the water must be considered in attaining maximum kayak speed.

The leg action stabilizes the hip joints, so the analysis is started at the hips, progressing upward toward the hands with each side done separately. The hands are pulling and pushing with a constantly changing force; thus, these external forces must be measured and applied at each instantaneous position analyzed. A flat strain gauge placed between the hand and the paddle would be required to obtain such measurements.

After the upper body is analyzed, the horizontal forces obtained must be applied to the legs at the hips, with the foot as the fixed point. The weight of the trunk is fully supported by the seat, and the legs are partially supported by the seat and footrest. The percentage of the leg weight to be used must be determined using the total leg center of gravity; then the force and moment analysis is completed. Measurements that may be obtained are the body seg-

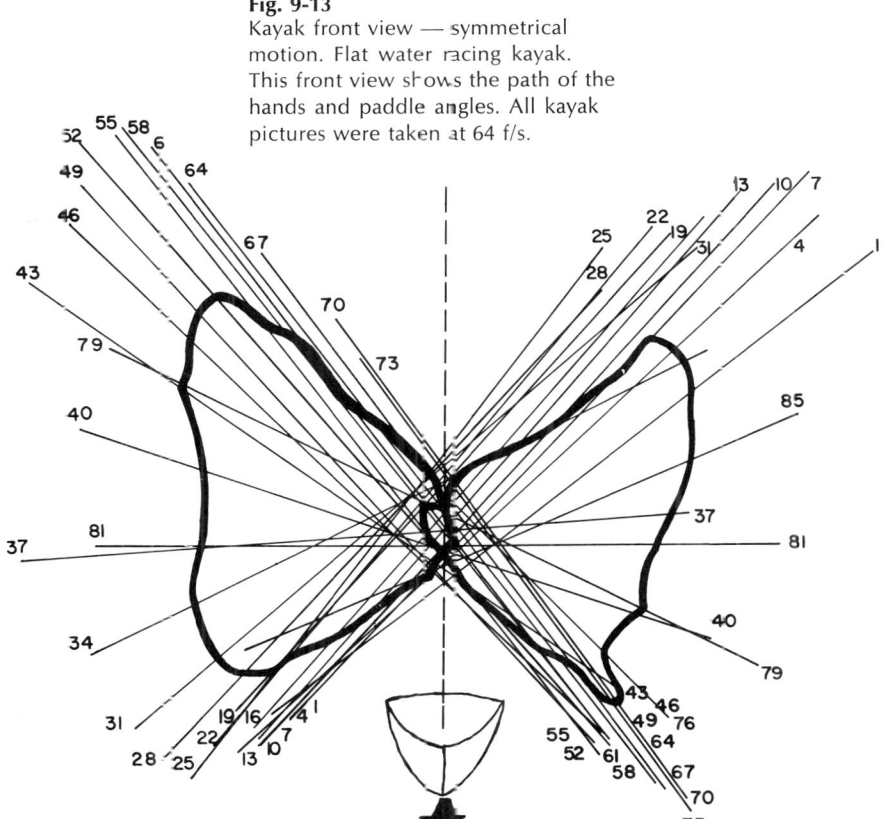

Fig. 9-13
Kayak front view — symmetrical motion. Flat water racing kayak. This front view shows the path of the hands and paddle angles. All kayak pictures were taken at 64 f/s.

ment positions, paths of the joint centers, length of stroke, relative movement of the hands during the simultaneous push-pull, paddle movement relative to the kayak and to the water, height of the seat, paddle angles, paddle pivot points, and the moments of force. As practice is usually seasonal, the information obtained may be used to determine a winter training program.

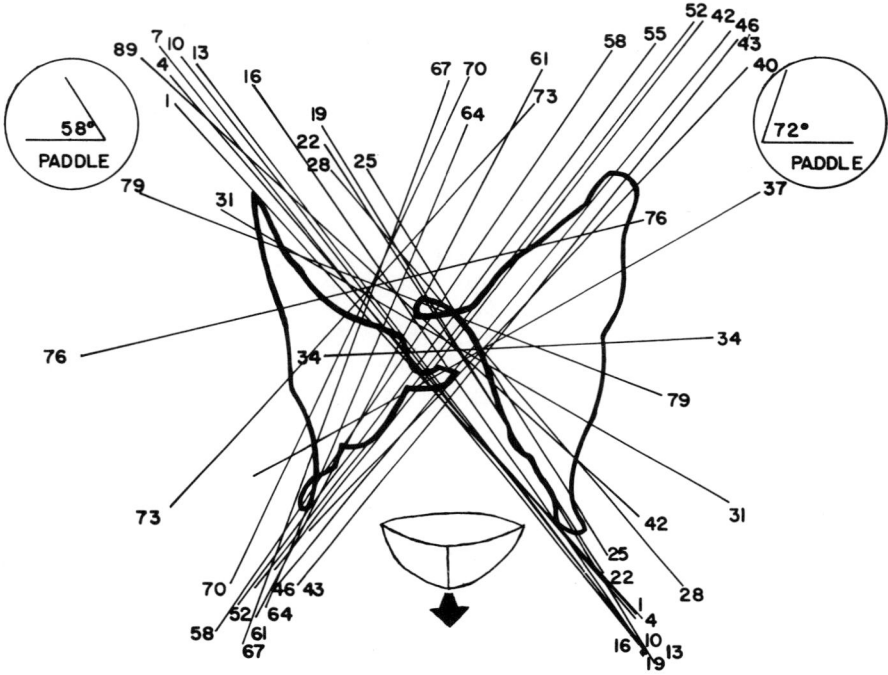

Fig. 9-14
Kayak front view — a non-symmetrical pattern resulting in differences in the paddle angles causing a zig-zag motion.

Selected tracings are presented here to point out variations in styles of paddling. Six Olympic and six good, but non-Olympic paddlers were analyzed. Figure 9-13 shows a symmetrical pattern of the hands from a front view, while in Fig. 9-14 a high left and a low right hand result in a 14° difference in maximum paddle angle. Figure 9-15 shows an exaggerated crossover of the left hand resulting in a wide sweep stroke, causing the boat to zigzag excessively. The front view is used in determining paddle angles, upper arm angles to the horizontal, and upper arm angles relative to the line of the shoulders (see Fig. 9-16).

The top view is very important for obtaining the symmetry of the motion pattern, the pivot point and positions of the paddle, the "rotation" of the shoulders about the spinal axis, and the angle of the upper arm to the line of the shoulders (Fig. 9-17). Figure 9-18 presents a symmetrical pattern, while

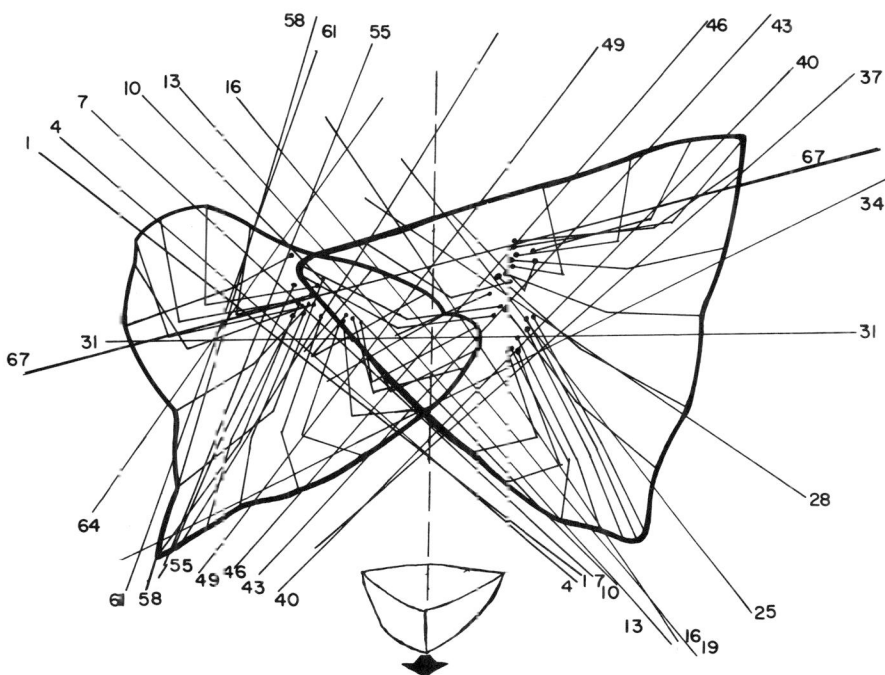

Fig. 9-15
Kayak front view — a hand pattern that resulted in an excessive zig-zag due to the exaggerated cross over of the left hand.

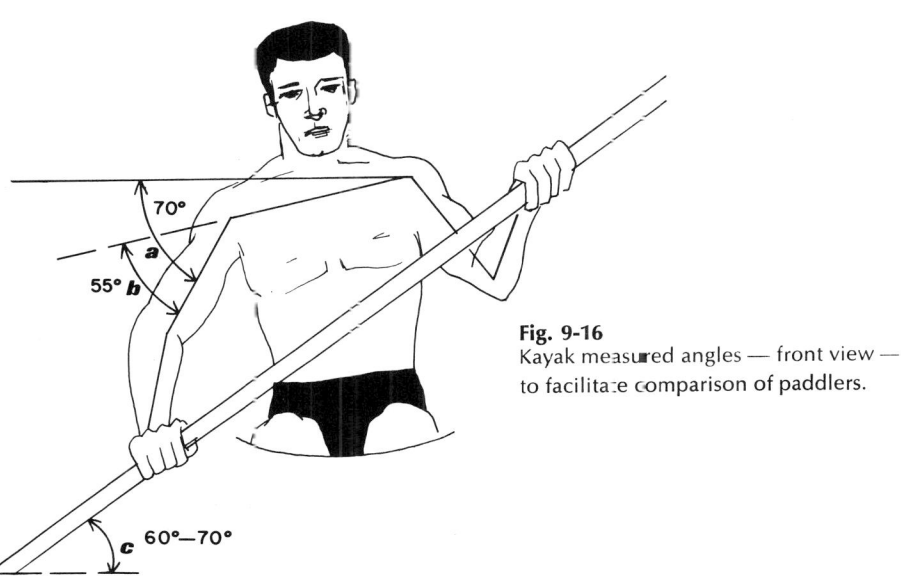

Fig. 9-16
Kayak measured angles — front view — to facilitate comparison of paddlers.

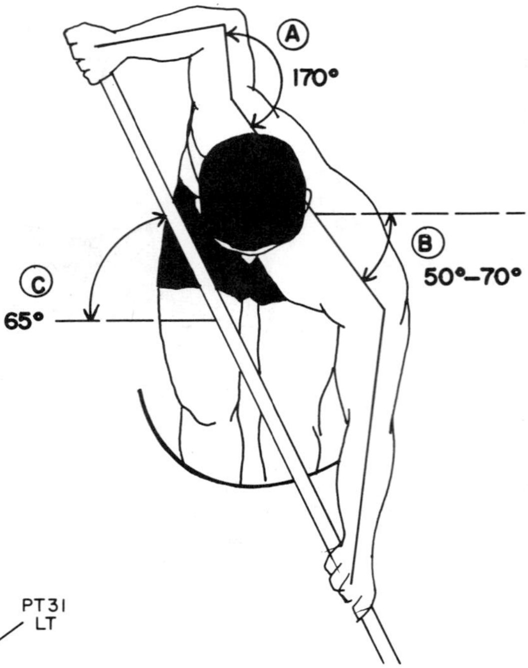

Analysis of Selected Sports

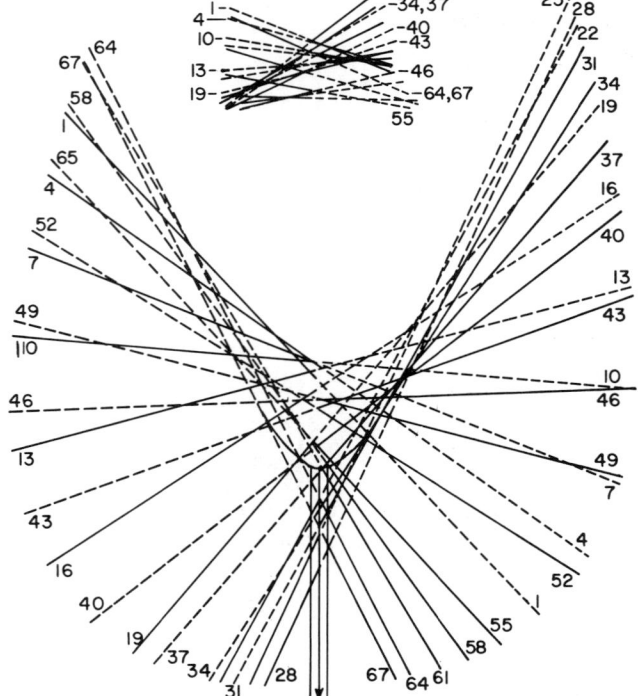

Fig. 9-17
Kayak measured angles — top view — to facilitate comparison of paddlers.

Fig. 9-18
Kayak — top view of paddle and shoulders showing a symmetrical motion (D. O'Keefe).

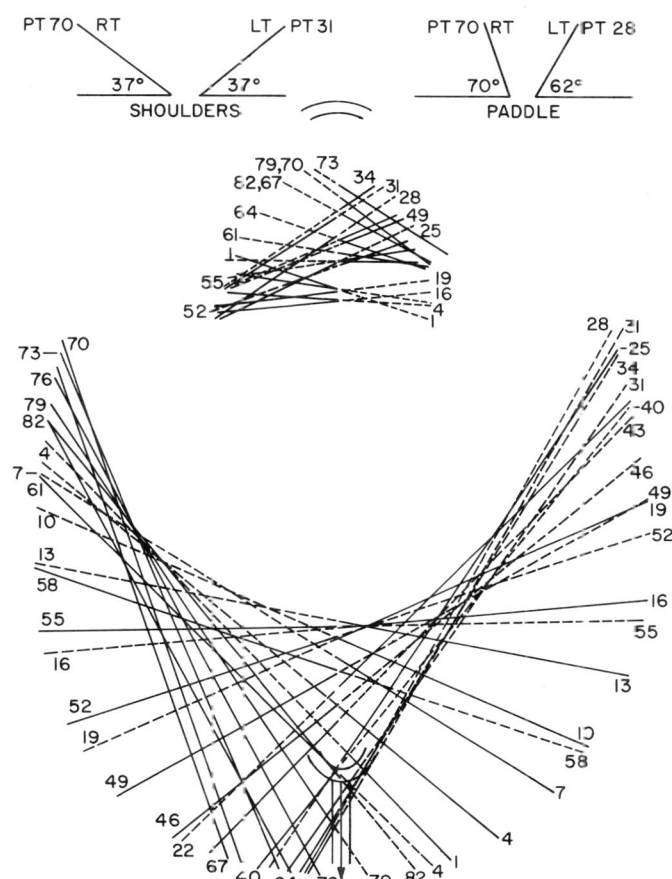

Fig. 9-19
Kayak — top view of the exaggerated left hand cross over also shown in Figure 9-15 (Olson).

Fig. 9-19 shows the paddler with the exaggerated left hand crossover (see Fig. 9-15). Figure 9-20 shows the paddler with a strong right pull resulting in an imbalance in the shoulder rotation, and a distinguishable paddle pivot point on the left side (circled).

The side view tracings are used to measure the trunk, hip, upper arm, and forearm angles during the pull and push (Fig. 9-21). Figures 9-22 and 9-23 show the right and left side tracings of a paddler with a strong right pull. This illustrates the relationship of the paths of the joint centers relative to each other, and the difference in trunk and arm positions causing nonsymmetry.

The kayaking analysis is aided by dividing the total stroke into three phases (Fig. 9-24a–d). *Phase 1* — from position 1 to 2 (paddle entry to vertical paddle); *phase 2* — from position 2 to 3 (vertical paddle to a position where the push arm is fully extended and horizontal); and *phase 3* — from position 3 to 4 (push arm horizontal to paddle exit). Optimum body positions were chosen by comparing the twelve paddlers measured and relating the joint moments of force and performance times. External force measurements were not possible, so an estimated maximum force was chosen for the perpendicular paddle position and

134

Analysis of Selected Sports

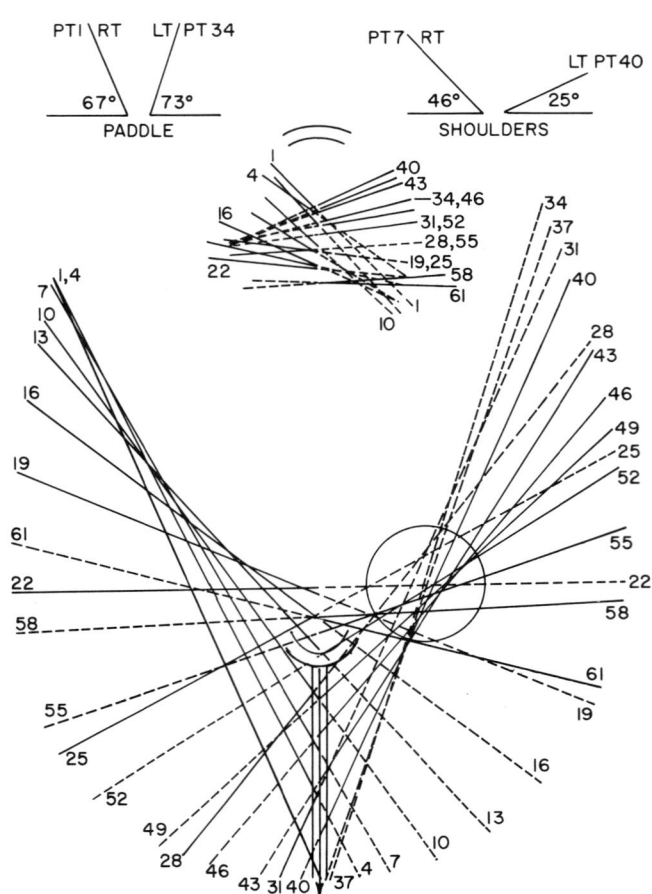

Fig. 9-20
Kayak — top view of a paddler (Pickett) with a strong right pull showing a distinguishable paddle pivot point on the left side only (circle).

Fig. 9-21
Kayak — side view angles measured to facilitate comparison of paddlers.

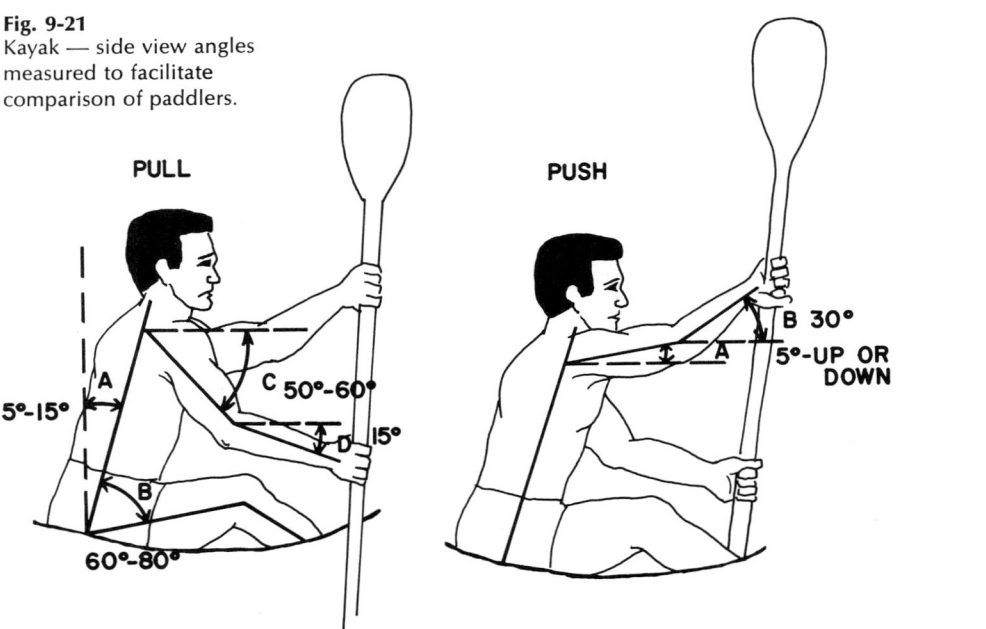

Analysis of Selected Sports

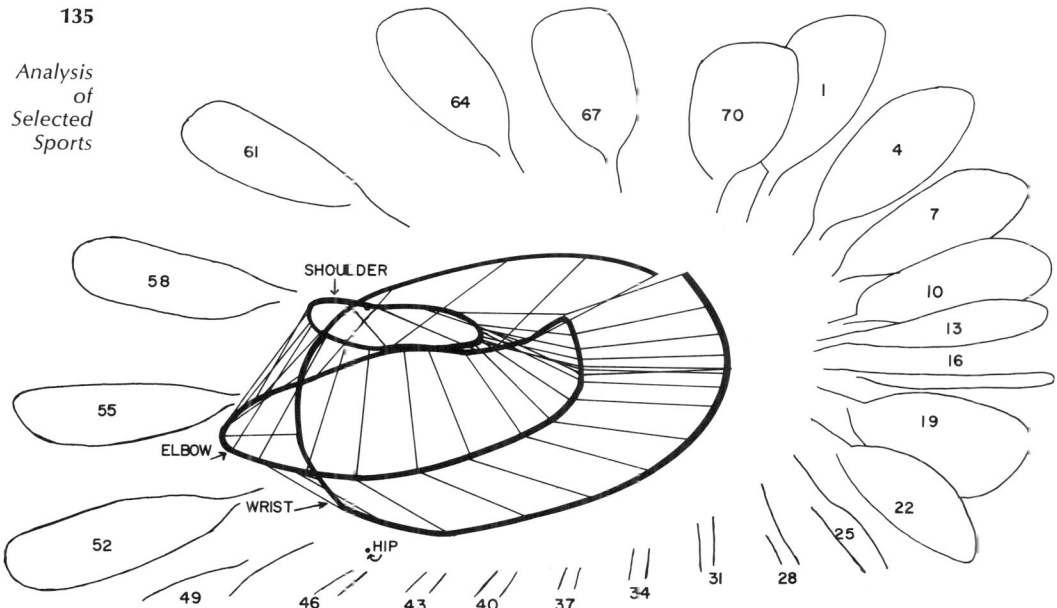

Fig. 9-22
Right side view of Olympic paddler John Pickett.

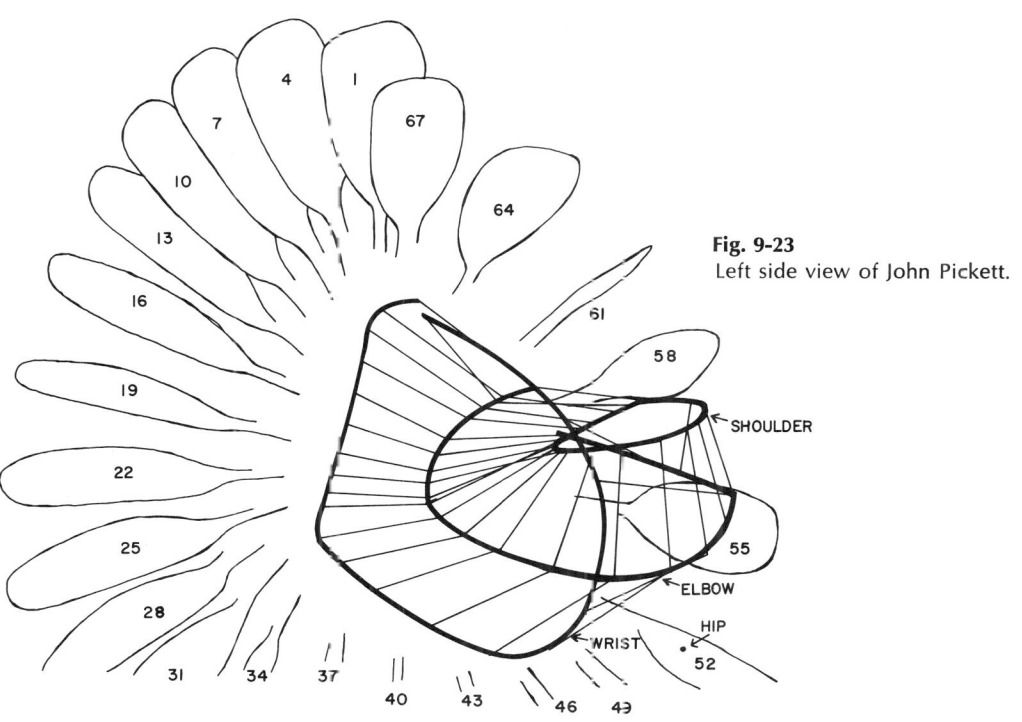

Fig. 9-23
Left side view of John Pickett.

136

Analysis of Selected Sports

POSITION 1

POSITION 2

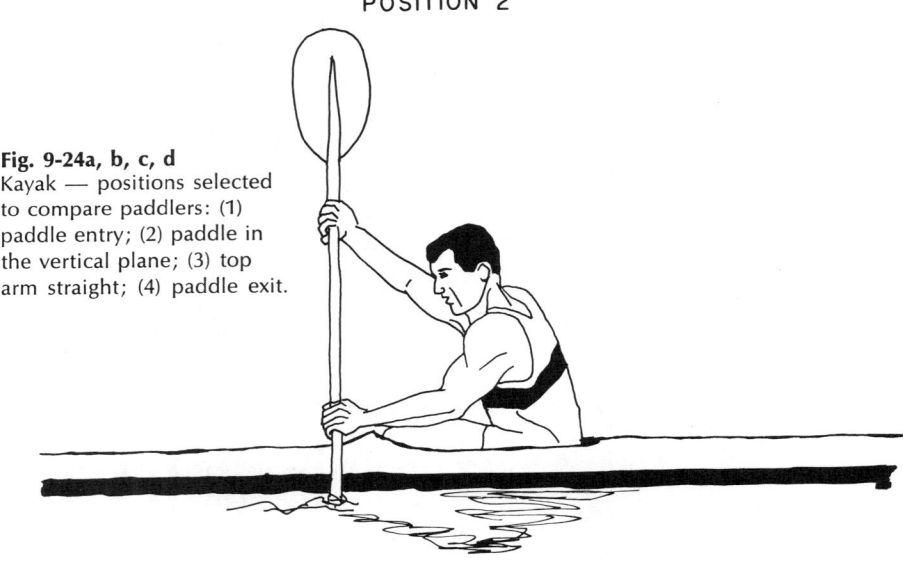

Fig. 9-24a, b, c, d
Kayak — positions selected to compare paddlers: (1) paddle entry; (2) paddle in the vertical plane; (3) top arm straight; (4) paddle exit.

POSITION 3

POSITION 4

decreased for each instantaneous position according to the paddle angle. The following conclusions were obtained from the composite tracings and moment analyses.

1. The hand and forearm should remain in alignment throughout the stroke.
2. A symmetrical motion produces lower moments of force at all joints.
3. The arm muscle requirements increase as the forearm angle increases from the horizontal (side view). It is recommended that the forearm angle to the horizontal not exceed 15° downward during the pull or 30° upward during the push when the paddle is in the vertical position (position 2). This requires a high elbow during the push with the upper arm remaining very near the horizontal throughout the push. The upper arm angle during the pull is approximately 20° at paddle entry, and 50° to 60° downward at position 2. The paddle angle at entry varied from 24° to 55°, but was about 40° for most paddlers.
4. There is very little trunk flexion and extension, but a great deal of shoulder rotation. The shoulders turn as far as 70° from the mid-position for the Olympic paddlers (see Fig. 9-17). Slight flexion occurs during the first half of the stroke and extension occurs during the second half of the stroke.
5. The pumping action of the legs changes the knee angle from about 120° to 135°. Straightening of the knee occurs on the pull side. The combination of the trunk movement and alternate leg movement changes the hip angle from about 60° to 80°. The pull tends to move the hip to the front of the seat, so the push of the leg on the same side counters this force. Because the push tends to move the hip backward, the medial rotators of the opposite leg (same leg countering the pull) must counter this hip rotation. Therefore, a toe strap should be helpful for the push side.
6. The push hand moves a greater distance than the pull hand during phase 1, and the opposite is true during phase 2. Therefore, the push produces about 60% of the total movement at the beginning of the stroke, and the pull produces about 60% at the end of the stroke.
7. The paddle is in the water about 70% of the time, although the range was from 78% to 50%.
8. The paddle remains in the water at the place of entry during the whole stroke. A poor ending causes the kayak speed to move the paddle forward rela-

tive to the water. If paddle exit during feathering is done too slowly, an unwanted drag results.

9. The directions of the pull and push and the body segment positions are more limiting than are muscle forces. If the push crosses the midline before phase 3, greater muscle force is required to achieve equal kayak speed. Using all body segments reduces the magnitude of the joint moments, and so trunk rotation aids the arm action.

10. A motion pattern that shows an alternating use of a muscle group, rather than a sustained contraction, is best suited for endurance paddling.

11. The joint moment analysis shows that fast paddling does not require great muscular strength; therefore a training program should emphasize muscular endurance. The maximum moments at each joint were compared to weight training exercises. The following program lists exercises that are similar to paddling motions. Repetitions should be gradually increased using the listed weights.

   a. Alternate curls—25 lb dumbbells
   b. Incline press 45° — 30 lb dumbbells
   c. Supine press — 30 lb dumbbells
   d. Pulley latissimus (Lat. machine) — 50 lb
   e. Pulley triceps (Lat. machine) — 35 lb
   f. Swim cables (shock cord) or wall pulleys, alternate pulls facing the wall — 15 lb
   g. Swim cables, alternate pushes facing away from the wall — 15 lb
   h. Pulley, supine pull down (Lat. machine) — 50 lb
   i. Sit-ups, knees slightly bent, no weight
   j. One-half deep knee bend — 25 lb

Every sport should be explored by collecting data on as many athletes as possible. When this is done, information pertaining to the proper sequence of action of all the body segments will lead to an understanding of optimum conditions of performance.

TENNIS SERVICE

The tennis serve is used to illustrate the problems encountered in analyzing a throwing motion, and to compare the serves of the professional players Rod Laver and Ken Rosewall. The link system is started with the front foot, progressing up the leg, across the hip joints, and up through the racket arm. This eliminates the motion of the back leg and ball toss arm (Fig. 9-25 and 9-26).

Four problems must be resolved:

1. Locate the center of gravity and determine the moment of inertia of the combined hand-racket segment;
2. Determine the force of impact that is transmitted to the hand through the racket due to ball impact;
3. Determine the weights to be used for the separate pelvic-upper trunk motions; and
4. Analyze the arm and leg that are out of the link system separately, and apply the forces obtained to the link system at the appropriate joint.

A weight equal to the hand was attached to the racket. This system was swung about a point equivalent to the position of the wrist joint to determine the moment of inertia, and was also balanced to determine the center of gravity.

Information pertaining to the force of impact is unavailable, so the motion analysis does not include the changing muscle forces at the point of impact.

Because the shoulders are rotated a great deal relative to the hips, the pelvic area and upper trunk are considered as separate segments. The pelvis is 24.7% of the whole trunk, so the pelvic weight is used for the hip-to-hip segment, and 75.3% of the total trunk weight is used for the trunk segment. The hip-to-shoulder distance is used for the trunk length, so the angular motion of this segment is a combination of trunk flexion-extension and shoulder rotation.

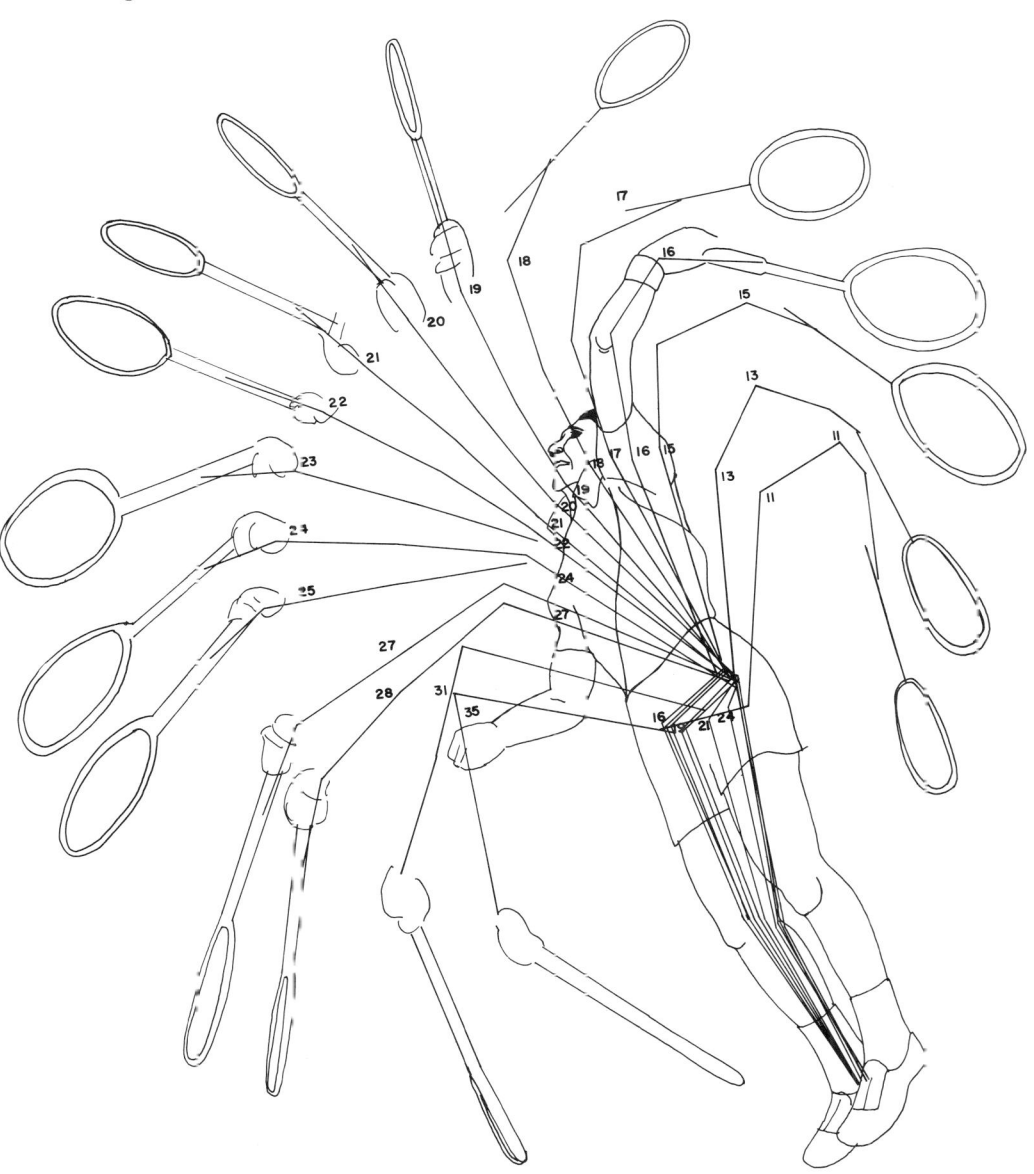

Fig. 9-25
Tennis service — Rod Laver.

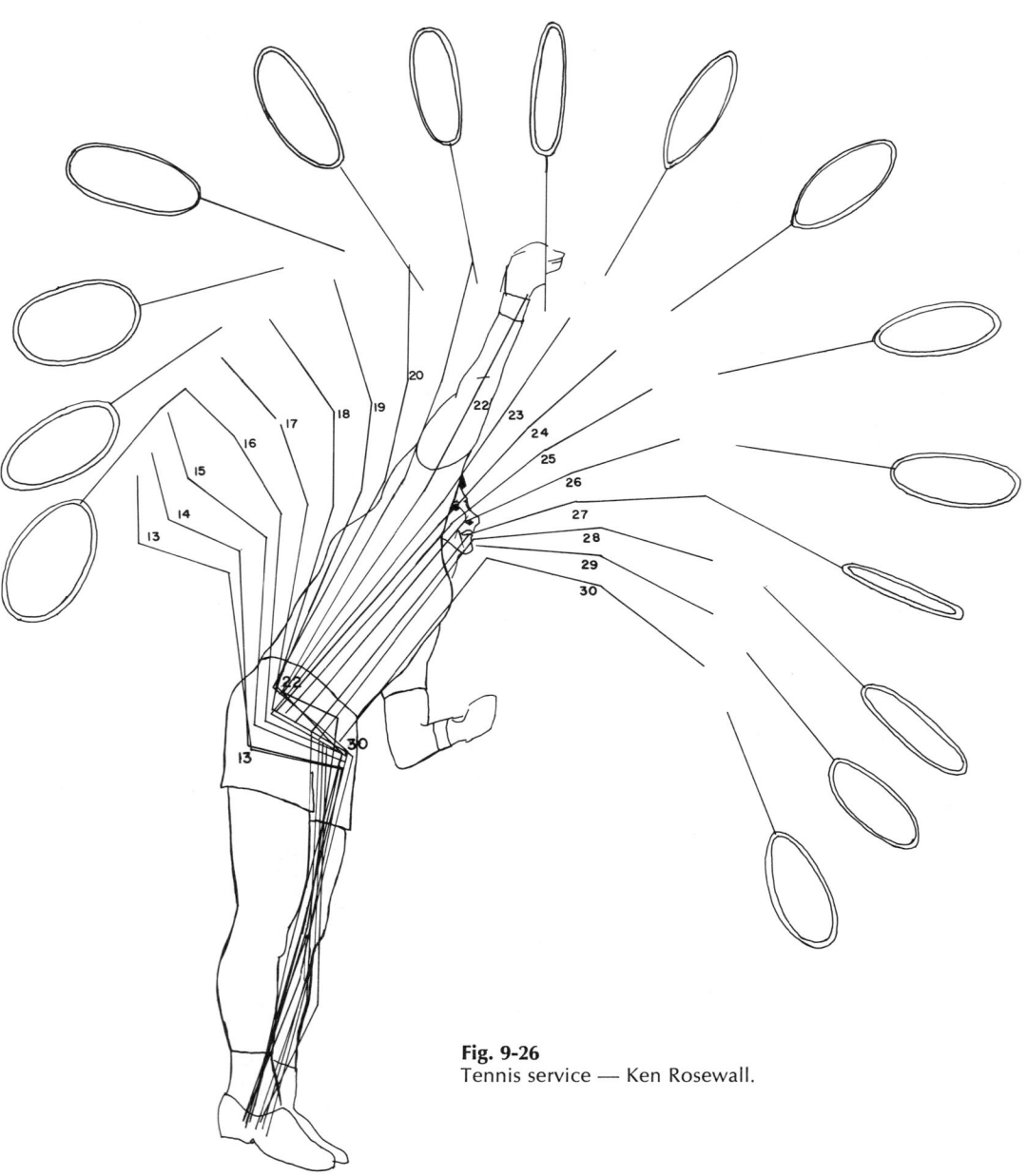

Fig. 9-26
Tennis service — Ken Rosewall.

At the time the racket has started its forward swing, the arm and leg outside the link system have finished their major contribution. These body segments were not included in the analysis of Laver and Rosewall for this reason. When segments outside of a chosen link system play a major role in the motion, they must be analyzed and included in the total analysis (e.g., in the running broad jump or high-jump take-off).

A comparison of the serves of Laver and Rosewall is presented to illustrate the use of the computer program output (Appendix B) for analyzing the motions. The analysis compares their differences relative to attaining maximum racket head velocity and maximum ball velocity. A comparison of the absolute racket angular velocity at impact shows that Laver's racket was moving 34.7 rad/sec while Rosewall's was moving 20.3 rad/sec. Rosewall's racket was moving 30 rad/sec two positions before impact while Laver's maximum velocity occurred at impact.

The absolute maximum deceleration of a body segment increases the velocity of the next segment. A comparison of maximum decelerations shows that both players decelerate the shank early. Laver then has the thigh, trunk, upper arm and forearm decelerating in sequence before impact. Rosewall, however, uses the interaction between the trunk and upper arm well, while the thigh and hips play a minor role. A deceleration of the hand and racket slightly before impact is evidence of the preparation made in anticipation of impact. This pattern has been consistently found in all motions involving impact (in the judo back-fall, karate and boxing blows, and golf, hardball, and squash impacts).

The section of the program which zeros the angular velocities and accelerations of each body segment in turn to indicate relative motion contribution, shows that the trunk is the dominant segment just before and through impact for Laver. In comparison, Rosewall's legs and hand are most influential up to impact, with the trunk playing a secondary role.

The joint moments of force show that the leg and hip moments are the same before impact for both players, even though the thigh and trunk motion pattern differences are discernable from the illustrations. The different motion patterns resulted in Rosewall using more muscle for upper arm and elbow extension, while Laver used slightly more muscular force for hand flexion. Rosewall's stopping action during the follow-through was more abrupt, resulting in greater moments of force at all joints.

Although Rosewall's ball velocity was slower than Laver's after impact, the former's ball velocity relative to racket velocity before and after impact was greater. This indicates that his striking mass was greater, which is a result of grip firmness. For these two serves only, the computer output shows that Laver's sequence of action between the body segments produced a faster ball velocity with less muscular effort.

It is necessary to analyze a large number of players to gain knowledge relevant to the optimum serving pattern, but the results would probably show that Laver's timed sequence of the body segments is close to perfection.

GOLF DRIVE

Golf presents a special problem because both hands are attached to the club, and the right and left sides of the body have a different motion pattern. As in the other sports discussed above, there is a great deal of useful information that may be obtained using the stop-action projector. Some measurements that may be made by stopping the action on individual frames are:

1. The width of stance and placement of the ball at address;
2. Time taken for the back swing, and the time to impact after the downswing begins;
3. Position of the arms during the backswing when the wrist "cock" begins;

Fig. 9-27a through e
Golf drive, side view — Sandra Haynie. (Haynie chosen for illustration because of her extremely long backswing.)

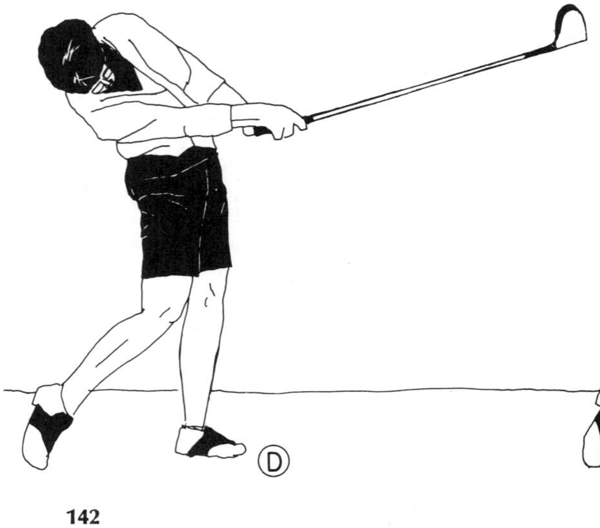

142

REAR VIEW

Fig. 9-28a through e
Golf drive, rear view — Sandra Haynie.

Fig. 9-29a through k
Golf drive — Tom Weiskopf. Weiskopf chosen for illustration because he represents a large group of golfers having a backswing stopping near the horizontal. His center of gravity changes are given relative to a dot reference point.

4. The position of the arms and club, and the angle between the forearm and club at the top of the swing;
5. Amount of club movement backward after the front leg starts its forward motion;
6. Maximum hip and shoulder rotation before the downswing begins;
7. Body and arm positions when the wrists begin to "uncock" on the downswing;
8. The body position at impact;
9. The club head velocity before and after impact, and ball velocity after impact (these three measurements are used to calculate the striking mass using the conservation of momentum formula);
10. Position at which the arms and hands start the roll-over after impact;
11. The length of the follow-through;
12. The instantaneous total body centers of gravity (this determines the shift of weight throughout the swing and allows a calculation of the percentage of total body weight on each foot).

Motion picture measurements of 20 touring professionals show that there are large variations in certain aspects of the swing. The width of the stance varies as much as ten inches, and the body and club positions at the top of the swing have many variations. The club moves as far back as 40° below the horizontal (Sandra Haynie), 20° below (Tom Weiskopf), parallel to the ground (Ray Floyd, Arnie Palmer, and the majority of golfers), or stops as early as 40° above the horizontal (Doug Sanders) (see Figs. 9-27, 9-28, 9-29, 9-30, 9-31, and 9-32).

The rear views show that the left arm angle varies from 35° above the horizontal (Gardner Dickinson) to 60° above the horizontal (Jack Nicklaus and Ray Floyd) at the top of the backswing. Most golfers are in the 45° to 55° range (see Fig. 9-28a). The combination of a large trunk rotation and a low arm angle are conducive to producing a hook, while a high arm and a shorter

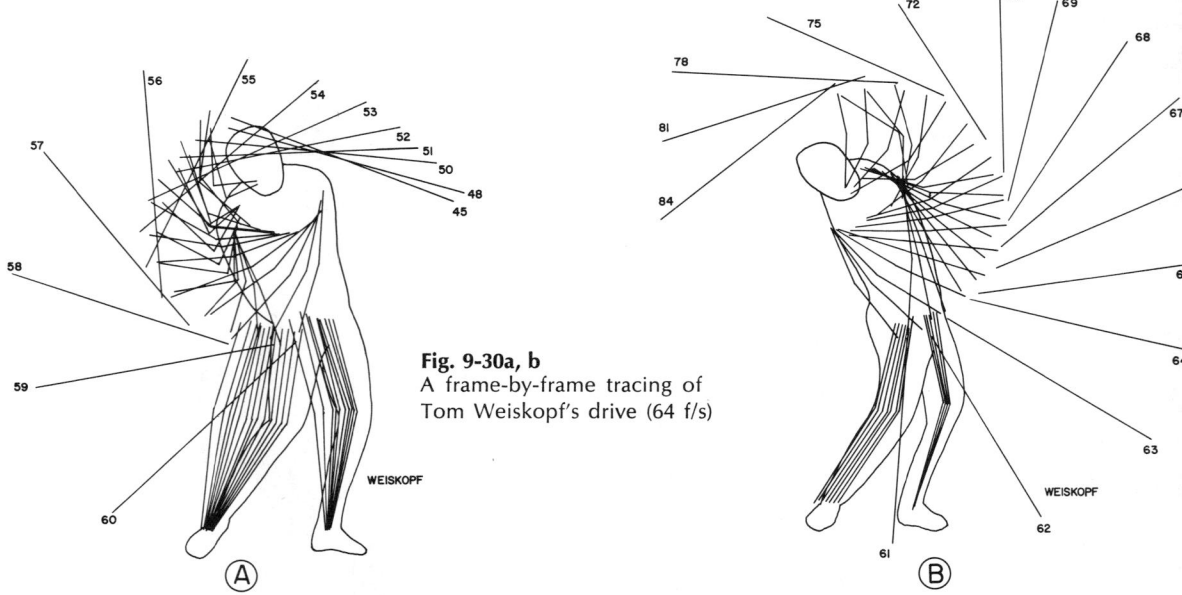

Fig. 9-30a, b
A frame-by-frame tracing of Tom Weiskopf's drive (64 f/s)

Fig. 9-31a through g
Golf drive, side view — Doug Sanders. Sanders was chosen for illustration because of his short backswing.

147

148

Analysis of Selected Sports

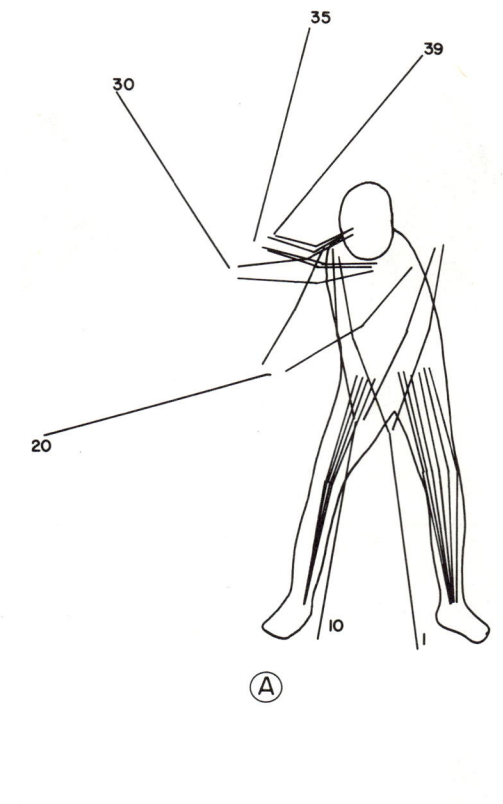

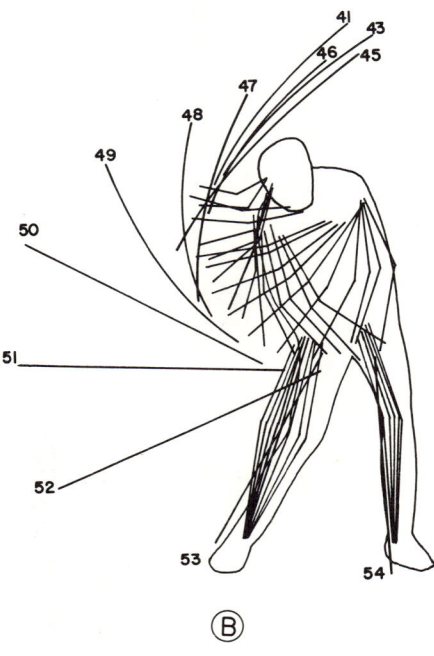

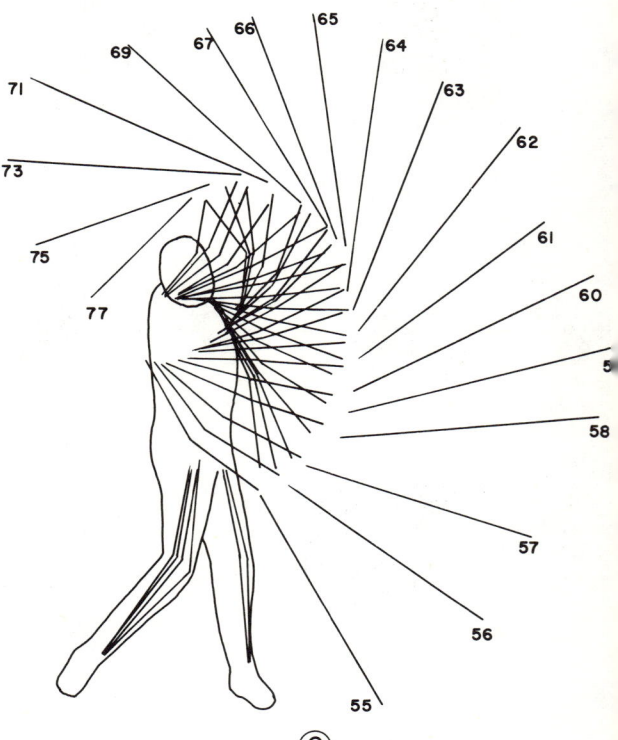

Fig. 9-32a, b, c
A frame-by-frame tracing of Doug Sander's drive (64 f/s).

trunk turn during the backswing is more likely to result in a slice, assuming a straight club head at impact.

The minimum forearm-club angle varies from 42° to 80°, with the majority ranging from 62° to 68°. The amount of hip rotation during the backswing varies, affecting the amount of left knee-bend and heel raise. The extremely long backswing of Haynie shows the hips rotating about 60° backward while the upper trunk (line through the shoulders) rotates about 110° (Figs. 9-27a and 9-28a). Most golfers rotate the hips about 30° to 40° from ball address, with the shoulders rotating about 90° around the spinal axis. This allows the left heel to remain on or close to the ground. Whatever the degree of hip and shoulder movement, it appears that all golfers rotate the shoulders relative to the hips maximally. The angle difference between the hips and shoulders should be measured from the top view (see Fig. 9-33).

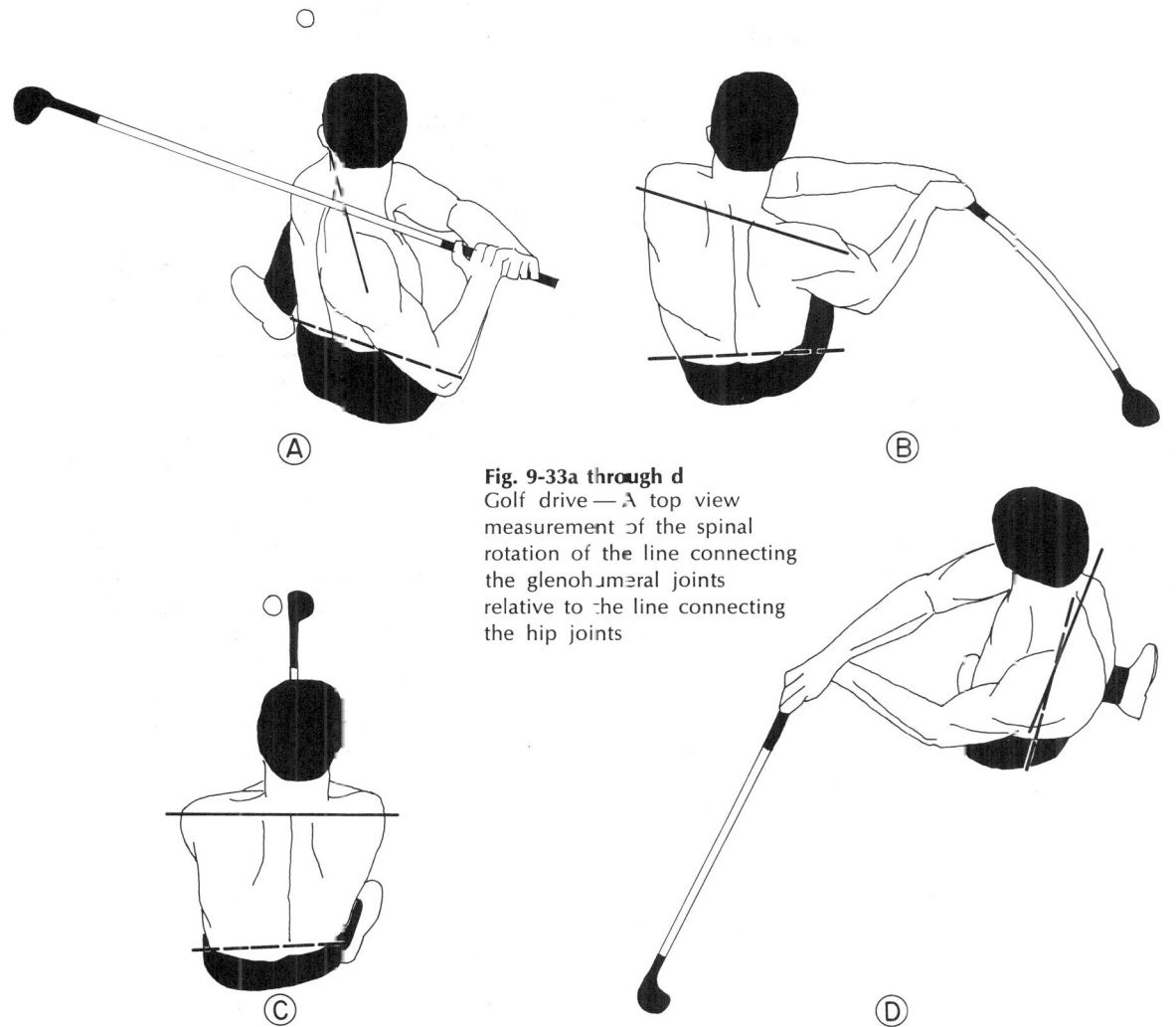

Fig. 9-33a through d
Golf drive — A top view measurement of the spinal rotation of the line connecting the glenohumeral joints relative to the line connecting the hip joints

All golfers start the shift of weight onto the front foot before the club has reached its maximum back position (see Fig. 9-29c and d). This timed body motion helps close the forearm-club angle with less muscular effort of the forearms. The angle between the club and forearm is held until the left arm is parallel to the ground, before the angle increases appreciably (see Figs. 9-27b, 9-29e, and 9-31b). Therefore, one of the most important functions of the hands is to hold the wrist "cock" at the beginning of the downswing. The ability to hold this club position until this last link in the system is ready for use appears to be one of the major differences between the good and the average player.

The position at impact is not the same as the position of address, due to the leg and trunk action. Most golfers have the hips turned open toward the line of ball flight more than the shoulders are at impact (compare Figs. 9-27c, 9-29g, and 9-31d).

The club moves to the positions shown in Figs. 9-27d, 9-29h, and 9-31e after impact, when the roll of the forearms produces a crossing of the wrists. The length of the follow-through depends upon the amount of trunk rotation, arm movement, and elbow bend (see Figs. 9-28e, 9-29k, and 9-31g).

The path of the total body center of gravity shows that there is only a slight backward movement during the backswing, but that the weight is shifted forward to a position close to the left heel at the beginning of the downswing (Fig. 9-29e). This rapid transfer of weight and an open body position at impact usually result in the right heel lifting off the ground while the right knee bends inward. The amount of weight on each foot is proportional to the distance of the center of gravity from each foot. (If the center of gravity is $\frac{1}{3}$ the distance between the feet from the left foot, the left foot is supporting $\frac{2}{3}$ of the weight).

The measurements of club head velocity before and after impact and of ball velocity to obtain the effect of grip firmness, show that the striking mass varies considerably. This points out the importance of the grip and the fact that the best swing is not the one that produces the fastest club head velocity, but rather the swing which allows a controlled impact to take place. A controlled impact has the club face in the desired position, the swing direction as desired, and the combination of grip firmness and club head velocity producing the maximum ball velocity.

A force and moment analysis requires that the swing be analyzed using three different link systems. One system starts at the left foot, moving up the left side through the left arm, while another starts at the right foot and moves up the entire right side. The third link system starts at the left foot, progresses up to the left hip, moves across to the right hip, and up the trunk through the right arm to the hand and club. (The same link is used in tennis, handball, squash, baseball batting and throwing, and the javelin, shot, and discus.)

Sanders and Weiskopf were compared by Gearon (I-34) because of their contrasting styles. He analyzed the right and left sides separately to obtain the following information:

1. Weiskopf attained the greatest angular velocity of the club, club head velocity, and ball velocity.
2. The contribution of the body segments due to maximum absolute decelerations stresses the importance of the right thigh, trunk, and forearm, and the left thigh and upper arm.

3. The contributions of the body segments due to the relative motion of each, stressed the importance of both legs, the trunk, the line of the shoulders spinal rotation, the left upper arm, and the right forearm and hand.
4. Weiskopf had greater maximum moments of force due to trunk rotation and arm motion, while Sanders' maximum moments were larger in the legs in both the clockwise and counterclockwise direction.
5. The relative difference between club head speed and ball velocity showed that Sanders' grip firmness produced a greater striking mass.

A joint moment analysis using the third link system substantiated the above points, and emphasized the importance of shoulder rotation relative to the hips. A general description of the swing to achieve maximum club head velocity using the force and moment data is as follows: After the preparatory backswing is near completion, the weight is shifted forward with a push from the right foot. The timed deceleration of the body segments starts with the left leg and progresses to the hip rotation, shoulder rotation, left upper arm, and right forearm. The left side is stabilized early in the swing so that the hip and shoulder rotation takes place about a pivot point near the left side. The hands hold the wrist "cock" until the right hand flexion is used to aid the angular club velocity, and then the firmness of the grip increases the striking mass to produce maximum ball velocity. All of the body segments aid the deceleration of the swing during the follow-through to reduce the club velocity gradually.

Squeezing the hands during impact becomes important only if the ball is contacted off center toward the toe. For a normally hit ball there should be very little or no twisting of the club in the hands. During the approximate $\frac{1}{1000}$ sec of impact, the club tends to rotate about a pivot point at the base of the right index finger (see Fig. 7-2b). The left hand is the counter to the torque and the right hand supports the pivot point. The readiness of the hands to perform this function must be timed so that the total swing also produces maximum club head velocity. The horizontal forces transferred through the club to the hands are not available, so the forces due to impact are not considered in the total body analysis.

A force and moment analysis is needed of many more players using all of the woods and irons, with the top view included if possible. Recording the swing of a player with high-speed motion pictures, when that player is hitting well, could be useful for future comparison with a swing that is troublesome.

chapter 10

MAXIMUM JOINT MOMENTS

THE APPLICATION OF MAXIMUM JOINT MOMENTS

A listing of the maximum moments of force for a large number of sports and for exercises used in conditioning programs and rehabilitation should be helpful for several reasons. The total muscular strength of a group of muscles must be equal to the demands placed upon it by a given task's motion. The maximum joint moments for an activity indicate the level of the muscular needs. Therefore, if a sport's maximum joint moments are known, preseason training can be designed to match them. If injury occurs during the season, a measure of readiness for return to action is available; thus the demands of the sport will not be in excess of the body musculature and result in reinjury. Rehabilitation exercises are designed to return an individual to functional living. Whether it is walking again or going back to manual labor, the exercises must be adequate to fulfill individual needs. A moment analysis could also be used in business and industry to determine the most efficient way of performing a job or to design equipment for the most effortless use. This could be done by keeping the joint moments to a minimum as well as by having alternate joint moment patterns for work requiring muscular endurance.

Table 10-1 lists a large number of exercises and sports motion patterns. This list has been revised and expanded since its original publication in 1967 (G-41). It is readily apparent that fast motions produce the largest joint moments, but they generally occur for only a fraction of a second. It is possible to design a weight lifting program that will prepare an individual for participation in a certain sport, and to determine readiness for participation by matching the maximum joint moments of the sport and the weight exercises.

Maximum Joint Moments

Table 10-1 Maximum Joint Moments for Exercises and Sports Motion Patterns

kg/m	Ankle Flexion	kg/m	Ankle Extension
2.6	Bicycling 15 mph	0.6	Push up
2.9	Rowing 36/min	1.0	Hack lift 15 lb
4.0	Pulley triceps 35 lb	2.9	Bicycling 15 mph
6.3	Swim cables 15 lb	5.0	Lift 50 lb box
6.7	Deep knee bend 100 lb	5.9	One-arm curl 20 lb
8.4	Swim cables 21 lb	6.0	Leg press (two) 110 lb
10.4	Duck waddle	6.9	Raise on toes (two) 50 lb
34.0	Latissimus pull 50 lb	8.5	Raise on toes (two) 100 lb
47.0	Bench step up 20 in.	10.0	Two-arm curl 80 lb
50.0	Jumping jack	12.0	Squash forehand, front leg
53.3	Squash forehand, front leg	13.3	Hack lift 120 lb
79.0	Squat jump, back leg	25.1	Deep knee bend 100 lb
150.0	Tennis serve	28.6	Two-arm press 80 lb
155.0	Standing broad jump	38.6	Back handspring
160.0	Javelin throw	44.0	Bench step up 20 in.
200.0	Golf drive	44.5	Two-arm press 150 lb
210.0	Squat jump, front leg	50.0	Rowing 36/min
220.0	Soccer kick	58.6	Swim start
230.0	Sprinting, ground leg	60.0	Standing broad jump
270.0	Badminton smash	80.0	Basketball vertical jump
270.0	Pitching fast ball	80.0	Clean 100 lb
275.0	Fencer's lunge, front leg	100.0	Clean 150 lb
275.0	Lacrosse shot	110.0	Squat jump, back leg
300.0	Basketball vertical jump	170.0	Squat jump, front leg
300.0	Fencer's lunge, back leg	185.0	Fencer's lunge, front leg
330.0	Sprinting, 2.5 lb ankle weight, ground leg	200.0	Golf drive
400.0	Track start	240.0	Sprinting, ground leg
500.0	Hammer throw	240.0	Sprinting, 2.5 lb ankle weight, ground leg
		240.0	Jumping jack
		280.0	Lacrosse shot
		280.0	Badminton smash
		290.0	Fencer's lunge, back leg
		300.0	Tennis serve
		300.0	Pitching fast ball
		300.0	Javelin throw
		300.0	Soccer kick
		300.0	Football shoulder block
		400.0	Track start
		500.0	Hammer throw

kg/m	Knee Flexion	kg/m	Knee Extension
0.15	Dips	0.5	Swim crawl kick
0.38	Pull-ups	1.0	Swim cables 15 lb
0.66	Straddle press up	1.2	L Lever
0.7	Swim crawl kick	2.3	Leg extension 5 lb
1.0	Straight arm planche	4.3	Leg extension 15 lb
1.0	Arched dip	5.1	Push-ups
2.3	Two-leg curl 10 lb	6.3	Leg extension 25 lb
2.8	Lift 50 lb box	6.7	Swim cables 21 lb
3.3	Bicycling 15 mph	8.0	Sprint, free leg
3.5	Two-leg curl 20 lb	9.4	Pulley, triceps 35 lb
4.3	Two-leg curl 30 lb	10.0	Clean 100 lb
4.8	Two-arm curl 40 lb	12.0	Clean 150 lb
5.3	Two-leg curl 40 lb	14.5	Leg press 55 lb
5.3	Squash forehand	14.7	Pulley triceps 50 lb

Maximum Joint Moments

kg/m	Knee Flexion	kg/m	Knee Extension
5.3	Tennis serve	15.8	Bowling, front leg
7.3	Two-leg curl 60 lb	15.8	Leg extension 65 lb
7.3	Standing broad jump	16.0	Pull-up
9.0	Pull-up	16.9	Hack lift 60 lb
9.5	Two-arm curl 80 lb	20.0	Sprint, 2.5 lb ankle weight, free leg
10.9	Swim start	21.8	Deep knee bend 50 lb
13.0	Sprint, free leg	24.3	Hack lift 120 lb
15.9	Football punt	25.2	Deep knee bend 75 lb
17.0	Two-arm press 150 lb	27.0	Duck waddle
18.0	Sprint, 2.5 lb ankle weight, free leg	28.0	Bench step up 20 in.
20.0	Rowing 36/min	28.6	Deep knee bend 100 lb
26.5	Back handspring	29.8	Standing broad jump
30.0	Clean 100 lb	30.0	Squat jump, back leg
30.0	Sprint, ground leg	40.0	Bicycling 15 mph
31.0	Football-soccer toe kick	47.0	Squash forehand
39.0	Badminton smash	50.0	Jumping jack
40.0	Clean 150 lb	50.6	Back handspring
45.0	Basketball vertical jump	59.0	Squat jump, front leg
81.0	Fencer's lunge, front leg	75.0	Basketball vertical jump
85.0	Soccer kick, side approach	87.4	Swim start
85.0	Squat jump, front leg	90.0	Football punt
90.0	Fencer's lunge, back leg	95.0	Tennis serve
130.0	Golf drive	110.0	Golf drive, front leg
150.0	Jumping jack	115.0	Badminton smash
190.0	Lacrosse shot	120.0	Football-soccer toe kick
200.0	Javelin throw	120.0	Fencer's lunge, back leg
240.0	Pitching fast ball	125.0	Soccer kick, side approach
240.0	Track start	130.0	Fencer's lunge, front leg
250.0	Sprint, 2.5 lb ankle weight, ground leg	150.0	Javelin throw, front leg
300.0	Hammer throw	150.0	Golf drive, back leg
		190.0	Pitching fast ball, front leg
		190.0	Lacrosse shot, front leg
		200.0	Sprint, ground leg
		200.0	Sprint, 2.5 lb ankle weight, ground leg
		230.0	Track start
		240.0	Football shoulder block
		240.0	Hammer throw

kg/m	Hip Flexion	kg/m	Hip Extension
0.7	Swim, crawl kick	0.2	Pull-up
1.1	Squash forehand	0.4	Swim, crawl kick
1.6	Deep knee bend 50 lb	0.6	Dip
2.9	Rowing 36/min	3.1	Straddle press up
3.0	Front giant	3.4	Arched dip
3.5	Knees to chest, hanging	4.4	Straight arm planche
3.7	Cast, parallel bars	4.4	Two arm press 80 lb
4.3	Bench step up 20 in.	4.6	Bowling, front leg
5.0	Squat jump, back leg	4.8	Two-arm curl 40 lb
5.1	L Lever	4.8	Two-arm, forward raise 50 lb
5.5	Peach basket	5.4	Swim cables, pull down 21 lb
5.5	Bicycling 15 mph	5.7	Back extension, hands on low back
6.0	Push-up		
6.0	Sit-up	5.8	Two-arm press 100 lb
6.6	V sit	6.4	Back extension, hands behind neck
6.8	Sit-up 10 lb		
7.8	Sit-up 20 lb	9.0	Jumping jack
9.0	Leg raise	9.3	Two-arm curl 80 lb
10.0	Pulley, triceps 35 lb	10.0	Pull-up

Maximum Joint Moments

kg/m	Hip Flexion	kg/m	Hip Extension
10.0	Head spring	10.0	Duck waddle
11.0	Standing broad jump	10.5	Front giant
11.3	Pulley, latissimus pull 50 lb	12.0	Cast, parallel bars
11.3	Swim cables, pull down 15 lb	12.0	Deep knee bend 25 lb
16.7	Peach to handstand	12.9	Deep knee bend 50 lb
20.0	Clean 100 lb	13.7	Peach to a handstand
22.0	Karate, downward blow	14.0	Bicycling 15 mph
22.0	Basketball vertical jump	14.6	Hack lift 60 lb
22.6	Squat jump, front leg	15.1	Head spring
25.0	Sprint, free leg	15.8	Deep knee bend 75 lb
26.0	Pull-up	16.7	Deep knee bend 100 lb
30.0	Football shoulder block	17.0	Peach basket
44.0	Badminton smash	18.0	Running broad jump, lift leg
49.0	Tennis serve	19.6	Standing broad jump
50.0	Swim start	21.0	Karate, downward blow
50.0	Golf drive, front leg	22.2	Hack lift 120 lb
52.0	Volleyball spike	24.0	Lift box 50 lb
54.0	Sprint, 2.5 lb ankle weight, free leg	24.0	Squash forehand, front leg
56.0	Running broad jump, free leg	24.0	Leg press 55 lb
60.0	Track start	25.0	Bench step up 20 in.
62.0	Sprint, ground leg	25.0	Sprint, free leg
69.0	Back handspring	25.0	Fencer's lunge, front leg
70.0	Golf drive, back leg	25.0	Squat jump, front leg
70.0	Sprint, 2.5 lb ankle weight, ground leg	26.0	Swim start
75.0	Fencer's lunge, front leg	26.0	Badminton smash
75.0	Football-soccer punt	27.0	Volleyball spike
90.0	Soccer, instep, side approach	30.0	Basketball, vertical jump
96.0	Javelin throw	31.6	Deep knee bend 150 lb
100.0	Football-soccer toe kick	32.0	Sprint, 2.5 lb ankle weight, free leg
110.0	Pitching fast ball, front leg	34.6	Back handspring
120.0	Lacrosse shot, overhand	35.0	Football-soccer punt
150.0	Hammer throw	39.0	Football-soccer toe kick
		44.6	Rowing 36/min
		45.0	Squat jump, back leg
		47.0	Sprint, ground leg
		60.0	Golf drive, both legs
		63.0	Fencer's lunge, back leg
		70.0	Tennis serve, front leg
		85.5	Clean 100 lb
		90.0	Lacrosse shot
		92.0	Soccer instep, side approach
		100.0	Javelin throw
		115.0	Clean 150 lb
		115.0	Track start
		125.0	Football shoulder block
		150.0	Pitching fast ball, front leg
		200.0	Hammer throw

kg/m	Back Extension (5th Lumbar)
1.5	Dip
3.6	Forward raise 25 lb
4.5	Arched dip
5.0	Back extension (no weight)
6.5	Two-arm press 80 lb
6.7	Two-arm curl 40 lb
8.0	Two-leg curl 120 lb
8.0	Pulley, supine pull down 70 lb
13.0	Deep knee bend 50 lb
13.4	Two arm curl 80 lb
20.0	Deep knee bend 100 lb
35.0	Lift large box 50 lb

Maximum Joint Moments

kg/m	Shoulder Flexion	kg/m	Shoulder Extension
1.4	Incline press 45°, 20 lb	0.5	Basketball free throw
2.2	Incline press 45°, 40 lb	1.2	Swim cables, pull down 15 lb
2.4	Incline press 18°, 20 lb	2.5	Pulley, supine pull down 50 lb
2.5	Basketball free throw	3.3	Standing broad jump
2.7	Swim crawl, Schollander	3.4	Tennis forehand
2.9	Squash forehand	3.5	Bowling, women
3.0	Incline press 45°, 60 lb	3.7	Pulley, supine pull down 70 lb
3.2	Basketball jump shot	3.8	Back handspring
3.3	Standing broad jump	3.9	Batting, back arm
3.9	Two-arm curl 30 lb	4.1	Basketball hook shot
4.1	Basketball hook shot	4.1	Basketball jump shot
4.1	Incline press 18°, 40 lb	4.2	Swim crawl, Schollander
4.3	Two-arm forward raise 30 lb	6.0	Pulley, latissimus pull 35 lb
5.0	Lift box 50 lb	6.0	Bowling, men
5.5	Bowling, women	7.8	Pulley, latissimus pull 45 lb
5.8	Incline press 18°, 60 lb	8.6	Hack lift 60 lb, bar free
5.9	Cast, parallel bars	9.1	Swim cables, pull down 21 lb
5.9	Two-arm curl 40 lb	9.1	Spring cables, lateral pull 50 lb
6.0	Two-arm curl 50 lb	9.2	Volleyball spike
6.5	Basketball vertical jump	9.2	Golf drive, back arm
6.7	Knees to chest hanging	9.9	Pulley triceps 50 lb
7.1	Two-arm forward raise 50 lb	10.0	Swim crawl, sprint, Van Kennen
7.2	Golf drive, front arm	10.0	Golf drive, front arm
8.3	Push-up	10.0	L Lever
8.4	Bent arm straddle press-up	11.0	Cast parallel bars
8.5	Tennis forehand	12.4	Rowing 36/min
8.5	Back handspring	12.5	Knees to chest hanging
8.9	Supine press 80 lb	12.6	Basketball vertical jump
9.0	Bowling, men	13.1	Swim start
9.1	Dip	17.0	Badminton smash
9.3	Swim crawl, sprint, Van Kennen	21.0	Clean 100 lb
9.8	Rowing 36/min	22.0	Lacrosse shot
9.9	Arched dip	22.0	Iron cross
10.0	Two-arm forward raise 70 lb	23.0	Handball serve
10.4	Straight arm-leg press-up	23.0	Football pass
10.8	Swim start	23.0	Squash forehand
10.8	Two-arm curl 100 lb	24.0	Tennis serve
11.0	Volleyball spike	25.0	Front giant swing
11.0	Supine press 100 lb	25.0	Javelin throw
11.0	Running broad jump, lift arm	26.0	Pull-up
11.6	Two-arm press 80 lb	28.0	Clean 150 lb
12.2	Straight body press-up	29.0	Peach to a handstand
14.0	Pull-up	29.0	Running broad jump, lift arm
14.4	Two-arm press 100 lb	30.0	Head spring
16.0	Supine press 150 lb	30.0	Pitching fast ball
16.6	Straight arm planche	45.0	Hammer throw
19.0	Golf drive, back arm		
20.0	Karate, downward blow		
20.0	Clean 150 lb		
21.0	Two-arm press 150 lb		
22.0	Badminton smash		
22.0	Batting, back arm		
22.0	Front giant swing		
25.0	Handball serve		
25.5	Javelin throw		
27.0	Tennis serve		
28.0	Lacrosse shot		
33.0	Headspring		
40.0	Peach to a handstand		
79.0	Pitching fast ball		
100.0	Hammer throw		

Maximum Joint Moments

kg/m	Elbow Flexion	kg/m	Elbow Extension
0.16	Bent arm straddle press	0.5	Two-arm press 80 lb
0.18	Swim cables 15 lb	0.5	Push-up
0.65	Two-arm press 100 lb	0.9	Incline press 18°, 40 lb
0.87	Push-up	0.9	Rowing 36/min
1.1	Basketball set shot	1.3	Standing broad jump
1.3	Football pass	1.6	Incline press 45°, 40 lb
1.5	Basketball jump shot	1.7	Swim crawl, Schollander
1.8	Standing broad jump	1.8	Supine press 80 lb
2.3	Two-arm forward raise 30 lb	2.3	Supine press 100 lb
2.5	Two-arm curl 30 lb	2.5	Tennis forehand
2.5	Swim crawl, Schollander	2.5	Incline press 18°, 60 lb
3.5	Pulley, supine pull down 50 lb	2.5	Knees to chest, hanging
3.9	Two-arm forward raise 50 lb	2.7	Basketball set shot
4.3	Two-arm curl 50 lb	2.9	Arched dip
4.5	Running broad jump lift arm	2.9	Hack lift, arms free 60 lb
4.9	Pulley, supine pull down 70 lb	3.0	Basketball jump shot
5.0	Lift box 50 lb	3.2	Incline press 45°, 80 lb
5.1	Karate, downward blow	3.5	Supine press 150 lb
5.1	Straight arm-leg press	3.5	Dip
5.2	Swim start	4.0	Basketball vertical jump
5.4	Two-arm forward raise 70 lb	4.2	Swim cables, pull down 21 lb
5.7	Cast to hands, parallel bars	4.5	Cast to hands, parallel bars
6.2	Badminton smash	4.6	Straight arm-leg press up
6.5	Golf drive, front arm	4.7	Swim start
6.6	Squash forehand	4.8	Incline press 45°, 120 lb
6.7	Swim crawl, sprint, Van Kennen	4.9	Football pass
6.8	Tennis forehand	5.0	Swim crawl, sprint, Van Kennen
7.0	Javelin throw	5.1	L Lever
8.0	Two-arm curl 100 lb	7.0	Badminton smash
9.0	Rowing 36/min	7.0	Running broad jump, lift arm
9.0	Golf drive, back arm	7.5	Squash forehand
9.1	Straight arm planche	9.0	Javelin throw
10.0	Volleyball spike	10.0	Pull-up
10.0	Handball serve	11.0	Pulley triceps 50 lb
11.0	Basketball vertical jump	11.5	Golf drive, back arm
12.0	Clean 100 lb	13.0	Golf drive, front arm
13.0	Two-arm curl 120 lb	14.0	Tennis serve
14.0	Batting, back arm	15.0	Karate, downward blow
15.0	Knees to chest, hanging	15.0	Clean 150 lb
16.0	Tennis serve	15.0	Handball serve
17.0	Clean 150 lb	17.0	Lacrosse shot
18.0	Lacrosse shot	17.0	Batting, back arm
25.0	Front giant swing	17.0	Volleyball spike
30.0	Head spring	24.0	Front giant swing
35.0	Pull-up	24.5	Headspring
50.0	Pitching fast ball	25.5	Pitching fast ball
90.0	Hammer throw	90.0	Hammer throw

kg/m	Wrist Flexion	kg/m	Wrist Extension
0.16	Straight arm-leg press up	0.03	Basketball set shot
0.36	Rowing 36/min	0.18	Swim crawl, Schollander
0.42	Basketball set shot	0.25	Basketball jump shot
0.44	Swim crawl, Schollander	0.58	Javelin throw
0.53	Two-arm curl 40 lb	0.62	Handball serve
0.6	Basketball jump shot	0.64	Swim crawl, Van Kennen
0.65	Wrist curl, two, 50 lb	0.65	Wrist extension 25 lb
0.82	Swim crawl, Van Kennen	1.0	Wrist extension 35 lb
0.86	Swim cables 21 lb	1.0	Volleyball spike
0.9	Wrist curl, two, 70 lb	1.1	Rowing 36/min
0.99	Javelin throw	1.2	Clean 100 lb

Maximum Joint Moments

kg/m	Wrist Flexion	kg/m	Wrist Extension
1.0	Two-arm curl 80 lb	1.3	Bent arm, straddle press up
1.3	Pulley triceps 50 lb	1.7	Tennis forehand
2.4	Badminton smash	1.8	Clean 150 lb
3.0	Clean 150 lb	2.3	Badminton smash
3.0	Pitching fast ball	5.0	Golf drive, back arm
3.1	Volleyball spike	6.1	Lacrosse shot
3.2	Handball serve	6.3	Tennis serve
3.4	L Lever	7.5	Golf drive, front arm
4.0	Tennis forehand	9.3	Pull-up
4.0	Knees to chest, hanging	12.5	Squash forehand
5.0	Golf drive, back arm	15.0	Knees to chest, hanging
5.2	Squash forehand	16.0	Batting, back arm
5.4	Tennis serve	20.0	Pitching fast ball
6.0	Lacrosse shot	22.0	Cast to hands
8.5	Batting, back arm	25.0	Peach to a handstand
18.2	Cast to hands	26.0	Headspring
24.0	Headspring	33.0	Front giant swing
27.5	Peach to a handstand	90.0	Hammer throw
30.0	Front giant swing		
35.0	Pull-up		
50.0	Hammer throw		

Two illustrations will show how the maximum moments may be used. A post-operative knee has been exercised, and the soccer player has advanced to a 65 lb leg extension, and a 40 lb two-leg curl. If the player should attempt a full swing instep kick, the knee moments of force would exceed the moments of the exercises and knee injury is likely. (The maximum leg extension moment from the exercise is 15.8 kilogram-meters, while the soccer kick is 125 kg/m and the leg curl maximum is 5.3 kg/m, while the knee flexion of the soccer kick is 85 kg/m.)

The second illustration deals with a low back muscle injury. A back extension exercise in the prone position may be painful due to the position or motion, or the maximum moments may be too great (hip extension 5.7 kg/m). Therefore, using the standing two-arm curl with 20 lb produces back extension moments of 2.4 kg/m while the trunk remains stationary and upright. (The moment listed for one arm using 20 lb is 4.8 kg/m.) When the curl weight is increased to 50 lb using both hands, the low back muscles are ready for the back extension exercise.

Table 10-1 lists all motions as flexion or extension, although this may not be the true motion. The moments are either clockwise or counterclockwise in the plane of the camera as the individual faces right or toward the camera. A golf swing has both upper arms moving in the counterclockwise direction with the right arm flexing and adducting while the left arm is extending and abducting. The simplification to flexion and extension is for convenience of interpretation; the motion should be related to the plus or minus direction for combined motions.

The study of kinesiology must ultimately result in joint moment analyses. The undergraduate course should begin the study of the forces due to motion, and the graduate course should continue with the determination of the magnitude of the forces and their relation to the joints and muscles. It is hoped that intensive work will be done by teams of analysts on each sport and on all the exercises used in rehabilitation and conditioning.

appendix A

PROBLEMS

1. You should know or review how to reduce trigonometric functions to functions with angles less than 90°. Here are some examples:

$$\cos 160° = -\cos 20°$$
$$\sin 200° = -\sin 20°$$
$$\cos 225° = -\cos 45°$$
$$\sin 345° = -\sin 15°$$
$$\cos 290° = \cos 70°$$

2. Determine the height x of the building shown in Fig. A-1.

   Solution:

   $$\tan 30° = \frac{x}{100 + y}$$

   $$.577(100 + y) = x$$

   $$\tan 45° = \frac{x}{y}$$

   $$1 = \frac{.577(100 + y)}{y}$$

   $$y = .577y - 57.7$$

   $$.423y = 57.7$$

   $$y = 136.4 \text{ ft}$$

   $$x = y$$

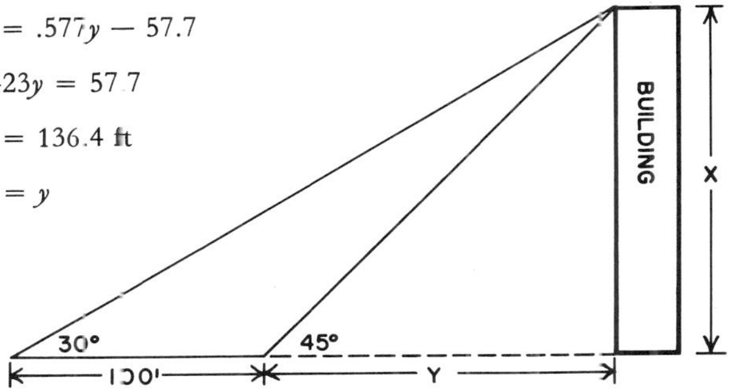

Fig. A-1
Building height.

3. Falling Body Formulas:

$$V = gt \qquad h = \tfrac{1}{2}gt^2 = \tfrac{1}{2}Vt$$
$$V_f = V_o + gt \qquad h = \tfrac{1}{2}gt^2 + V_o t$$
$$V_f = \sqrt{2gh}$$
$$V_f^2 = V_o^2 + 2gh$$

a. Given $V = gt$ and $h = \tfrac{1}{2}gt^2$. Complete the following table by calculating the missing values of V and h.

t	V	h
1 sec	32	16
2 sec	64	64
3 sec	96	144
4 sec		
5 sec		

b. If a ball is dropped from 100 ft, what is its velocity when striking the ground?

Solution:
$$h = \tfrac{1}{2}gt^2 \qquad V = gt$$
$$100 = 16.1 t^2 \qquad V = 32.2(2.5)$$
$$t = 2.5 \text{ sec} \qquad V = 80.5 \text{ ft/sec}$$

c. A ball is thrown straight up at an initial velocity of 100 ft/sec. What is the time and distance to the maximum height?

Solution:
$$V_f = V_o - gt \qquad V_o^2 = 2gh$$
$$0 = 100 - gt \qquad 100^2 = 64.4 h$$
$$\frac{100}{32.2} = t = 3.1 \text{ sec} \qquad h = 155.28 \text{ ft}$$

4. A 1 lb ball swings in a 10 ft arc (see Fig. A-2). What is the velocity at B? What is the velocity at B for a 300 cm radius? Note: (cm/sec)/30 = ft/sec.

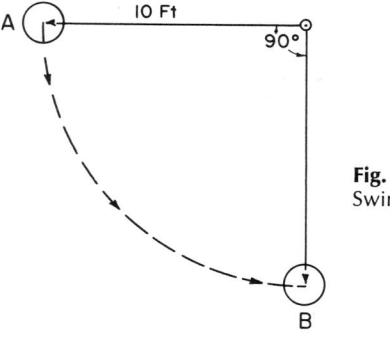

Fig. A-2
Swinging ball.

Solutions:
$$V = \sqrt{2gh} \qquad V = \sqrt{2(980)(300)}$$
$$V = \sqrt{2(32.2)(10)} \qquad = 766.8 \text{ cm/sec}$$
$$V = 25.4 \text{ ft/sec}$$

5. What is the resultant of the three forces shown in Fig. A-3?

Solution: $\sin 50° = \dfrac{y}{5}$ $y = 3.85$

$\cos 50° = \dfrac{x}{5}$ $x = 3.2$

$\sin 80° = \dfrac{y}{7}$ $y = -6.86$

$\cos 80° = \dfrac{x}{7}$ $x = 1.19$

Resultant:

y	x
3.85	3.2
−6.86	1.19
0	−10.0
−3.01	−5.61

Resultant vector $= \sqrt{x^2 + y^2}$
Resultant $= 6.37$ lbs

$\tan \theta$ in 3rd quadrant $= \dfrac{3.01}{5.61}$

$\theta = 28°$ or $208°$

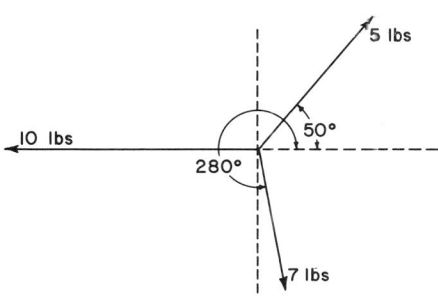

Fig. A-3
Resultant of forces.

6. If the coefficient of friction $\mu = .3$ in Fig. A-4, what weight must be hung at point A to keep the 100 lb weight from sliding?

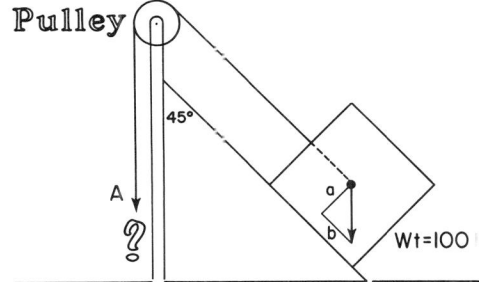

Fig. A-4
Friction on incline.

Solution: $\sin 45° = \dfrac{a}{100}$ $\mu = \dfrac{F}{N}$

$a = 70.7$ lb $.3 = \dfrac{F}{70.7}$

$b = 70.7$ lb $F = 21.2$ lb

$\ 70.7$
-21.2
$\ \overline{49.5}$ lb hung at point A

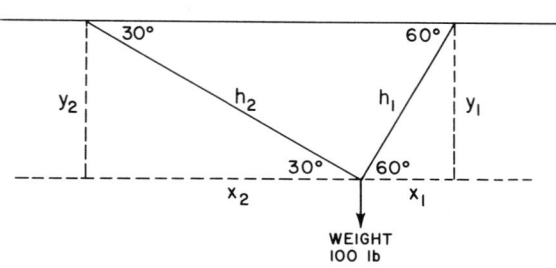

Fig. A-5
Weight on a string.

7. What is the force on strings h_1 and h_2 in Fig. A-5?

   Solution: $x_1 = x_2$

   $$\cos 60° = \frac{x_1}{h_1} \qquad \cos 30° = \frac{x_2}{h_2}$$

   $$h_1 \cos 60° = h_2 \cos 30°$$

   $$h_1 = 1.732 h_2$$

   $$y_1 + y_2 = 100$$

   $$\sin 60° = \frac{y_1}{h_1} \qquad \sin 30° = \frac{y_2}{h_2}$$

   $$h_1 \sin 60° + h_2 \sin 30° = 100$$

   $$h_2 = 50 \text{ lb} \quad \text{and} \quad h_1 = 86.6 \text{ lb}$$

8. Determine the initial velocity, time of flight, and range, given $\theta = 19°$ and $h = 6$ ft.

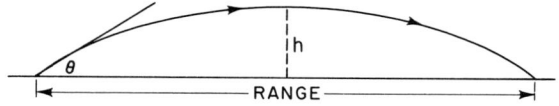

Fig. A-6
Range of trajectory.

   Solution:

   $$h = \frac{(V_o \sin \theta)^2}{2g}$$

   $$\sqrt{(6)(64.4)} = V_o \sin \theta$$

   $$V_o = 60.3 \text{ ft/sec}$$

   $$h = \tfrac{1}{2} g t^2 \qquad\qquad t = \frac{V_o \sin \theta}{g}$$

   $$\sqrt{\frac{6}{16.1}} = t \quad \text{or} \quad t = \frac{60.3(.33)}{32.2}$$

   $$.6 = t \qquad\qquad t = .6 \text{ to maximum height}$$

   Total time = 1.2 sec

   Range = $2 V_o \cos \theta t$

   Range = $2(60.3)(.945)(1.2)$

   Range = 136.8 ft

Appendix A

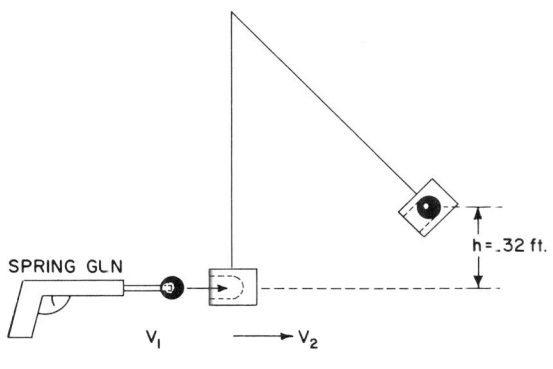

SPRING GUN

V_1 →V_2

$\frac{1}{2}(m_1 + m_2)V^2 = (m_1 + m_2)gh$

K.E. = P.E.

$V_2 = \sqrt{2gh}$

Fig. A-7
Conservation of momentum.

9. Combine a falling body problem with a conservation of momentum problem by solving for V_1 and V_2 in Fig. A-7. The velocities are given in the same terms, and the masses are given in the same terms. g would appear under both sides of the equation, so it can be eliminated. Given $h = .32$ ft, bullet weight = 5 g, and wood weight = 500 g. Then

$$V_2 = \sqrt{2gh} \qquad m_1V_1 = (m_1 + m_2)V_2$$
$$V_2 = \sqrt{(2)(32.2)(.32)} \qquad 5V_1 = 505(4.54)$$
$$V_2 = 4.54 \text{ ft/sec} \qquad V_1 = 458.5 \text{ ft/sec}$$

10. A 20 ft ladder is leaning against a wall with a man standing halfway up. Man + ladder = 200 lb. What is the horizontal force against the wall? Given: $\sin 60° = .866$, $\sin 30° = .5$, $\cos 30° = .866$, and $\cos 60° = .5$.

Solution 1:
$$\cos 60° = \frac{x}{10} \qquad \sin 60° = \frac{y}{20}$$
$$x = 5 \qquad y = 17.32$$
$$200(5) = 17.32x$$
$$x = 57.7 \text{ lb}$$

Solution 2:
$$\cos 60° = \frac{a}{200} \qquad \cos 30° = \frac{50}{x}$$
$$a = 100$$
$$100 \times 10 = b \times 20 \qquad x = \frac{50}{.866} = 57.7 \text{ lb}$$
$$b = 50$$

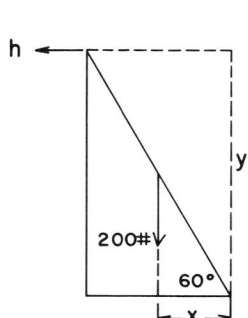

Fig. A-8
Ladder forces.

11. Figure A-9 illustrates a suspended wood problem. Hanging weights 10 lb–50 lb are reduced to the equivalent position of 60 lb (8 cm + 4 cm). P = contact point of wood and wall. Determine force h.

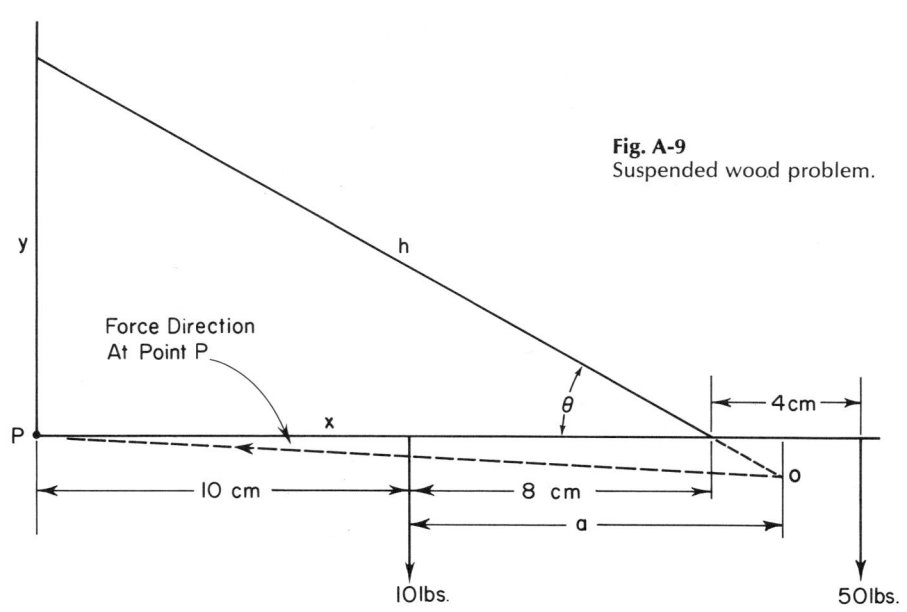

Fig. A-9
Suspended wood problem.

$$12(50) = 60a$$

$$x = 10 \quad \text{Position of 60 lb}$$

$$20(60) = 18x$$

$$x = 66.7 \text{ lb}$$

$$\sin \theta = \frac{66.7}{h}$$

$$h = 133.4 \text{ lb}$$

12. This friction problem combines determination of frictional force, the drawing of a free body diagram, and the resolution of three forces through a single point. A metal disc with a handle is free to slide on a metal plate. The string attached to the handle pulls on the disc until the sliding stops and the system is in static equilibrium (Fig. A-10a). The direction of the force between the metal plate and the disc is determined by finding the intersection of the weight and string lines of force. A line connecting the point of contact between metal and disc and this point of intersection of the weight and string produces the line of force desired. A free body diagram of the forces affecting the disc and handle shows that the force of friction is equal and opposite to the X component of the pull of the string (Fig. A-10b). The magnitude of the force of friction μ is determined by weighing the disc and measuring the angles of the lines of force (Fig. A-10c), and then solving the problem as follows.

Appendix A

Fig. A-10a, b, c
Sliding friction.

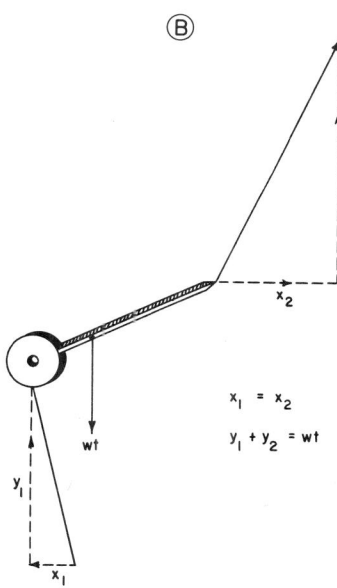

FREE BODY DIAGRAM

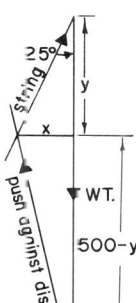

X = Force equal to friction between metal disc and plate beneath.

$$\tan 25° = \frac{x}{y} \qquad\qquad 500 - y = 335 \text{ g}$$

$$\tan 13° = \frac{x}{500 - y} \qquad\qquad \tan 25° = \frac{x}{165}$$

$$\tan 13°(500 - y) = \tan 25°y \qquad\qquad .467 = \frac{x}{165}$$

$$.23(500 - y) = .467y \qquad\qquad x = 77 \text{ g}$$

$$y = 165 \text{ g}$$

$$\mu = \frac{77}{500} = .154$$

13. Determine total body center of gravity for the body position illustrated in Fig. A-11.

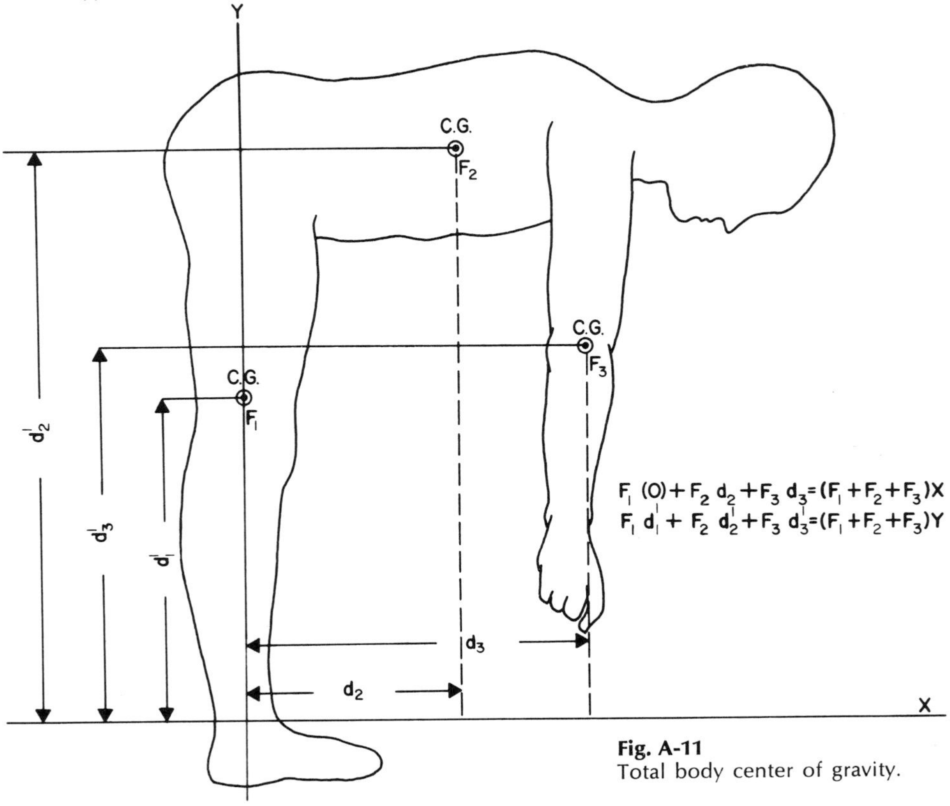

$$F_1(0) + F_2 d_2 + F_3 d_3 = (F_1 + F_2 + F_3)X$$
$$F_1 d_1' + F_2 d_2' + F_3 d_3' = (F_1 + F_2 + F_3)Y$$

Fig. A-11
Total body center of gravity.

14. Figure A-12 shows a collision diagram which we will use to study conservation of momentum. What is the direction and speed after collision if the cars stay together? The components of momentum are:

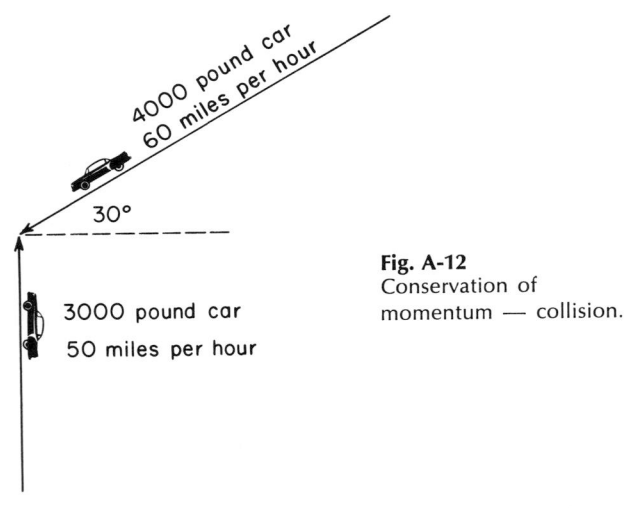

Fig. A-12
Conservation of momentum — collision.

$$\sin 30° = \frac{-y}{4000 \times 60} \qquad \cos 30° = \frac{-x}{4000 \times 60}$$

$$-y = 120,000 \qquad\qquad -x = 207,840$$

$$+y = 3000 \times 50 = 150,000$$

$$\text{resultant } +y = 150,000 - 120,000 = +30,000$$

$$\text{resultant } \quad x = -207,840$$

The direction after collision is then determined as follows:

$$\tan \theta = \frac{30,000}{207,840} = .1443 = 8°13' \text{ or } 171°47'$$

The resultant momentum is

$$\sqrt{207,840^2 + 30,000^2} = 209,994$$

Since

$$M_oM = mV \quad \text{and} \quad V = \frac{M_oM}{m}$$

then

$$V = \frac{209,994}{7000} = 30 \text{ mph}$$

15. Figure A-13 shows an apparatus used to study centripetal force. The horizontal force stretching the spring and resulting from the circular motion of the suspended mass (Fig. A-13a) can be calculated with this formula:

$$F = ma = \frac{mv^2}{r}$$

If the velocity of the spin is controlled so that the mass passes over the fixed pointer, determine the centripetal force using the following formulas and values

$$v = \frac{2\pi r n}{t} \qquad F = \frac{m}{r}\left(\frac{2\pi r n}{t}\right)^2 = \frac{4\pi^2 n^2 r m}{t^2}$$

Given: wt = 500 gms $n = 100$

$r = 20$ cm $t = 85$ sec

$$F = \frac{4(3.14)^2(100)^2(20)\frac{500}{980}}{(85)^2}$$

$$F = 556 \text{ gms}$$

where m = mass,
           r = radius of the path of the mass,
           n = number of revolutions (100 recommended), and
           t = time of n revolutions.

Determine the weight (see Fig. A-13b) that pulls the mass out to the fixed pointer, and compare it with the calculated force due to motion.

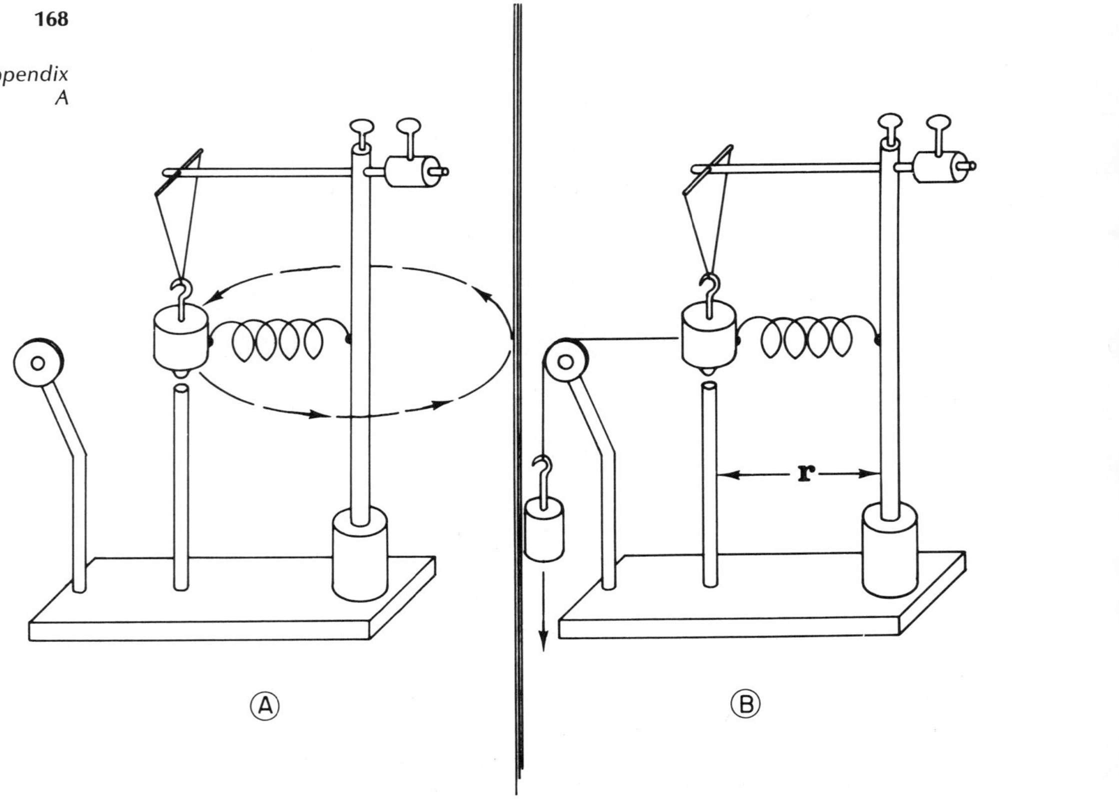

Appendix A

Fig. A-13a, b
Centripetal force.

appendix B

A COMPUTER PROGRAM FOR THE KINETIC ANALYSIS OF HUMAN MOTION

The following computer program will permit you to obtain the angular velocity and acceleration, forces, moments, center of gravity, and contribution of body segments in an analysis of a three-segment motion.

Two programs were presented in the *Journal of Biomechanics* in 1968 (D-9, Plagenhoef, Computer Programs for Obtaining Kinetic Data on Human Movement, copyright 1968 by Pergamon Press and reprinted by permission of publisher). These programs have been combined and expanded here so that minimal key punching produces:

1. absolute and relative angular velocities and accelerations with the data graphed for each segment,
2. the horizontal and vertical forces (graphed), and the resultant of these forces,
3. the joint moments of force (graphed),
4. the percentage contribution of each segment toward the total motion,
5. the total body center of gravity, and
6. the possibility of applying external forces due to impact.

All graduate students should photograph a three-segment motion and use this program to analyze the motion. One position should then be hand calculated using the relative motion method presented in Chapter 5 (see also Appendix C). The answers should then be compared with the computer output.

A symmetrical, two-handed, underhand basketball throw upward (Fig. B-1) is used to demonstrate the program; position 5 in Fig. B-1 is hand calculated in Appendix C to compare both the absolute and relative methods of analysis with the computer answers.

170

Appendix B

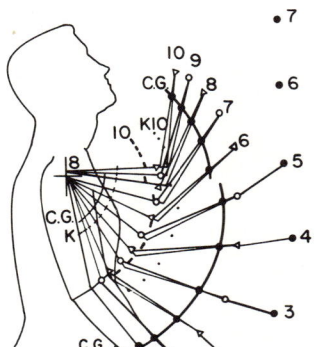

•10
•9
•8
•7
•6 JUST RELEASED
•5
•4
•3
•2
•1

Fig. B-1
Basketball vertical throw for computer program.

PUNCHED INPUT FOR BASKETBALL THROW (FIG. B-1)

```
   INPUT DATA   COMBINED PROGRAM
   BASKETBALL THROW   2 SEGMENTS-SEGMENT 2 FULLY KEYPUNCHED
 2
 11
  .0         .02
             2
  .436       .542
 28.6
 508
 2800.
 44
 5000.      50000.
-1.0+004  1.0+004-1.0+004  1.0+004-1.0+007  1.0+0071
-1.0+004  1.0+004-1.0+004  1.0+004-1.0+006  1.0+0062
 288.       28.6
 295.       28.6
 304.       28.6
 312.       28.6
 322.       28.6
 333.       28.6
 344.       28.6
 352.       28.6
 356.       28.6
 359.       28.6
 361.       28.6
 314.       28.6      2276.      22.4      26.6
 326.       28.6      2276.      22.4      26.6
 340.       28.6      2276.      22.4      26.6
 364.       28.6      2276.      22.4      26.6
 385.       28.6      2276.      22.4      26.6
 404.       28.6      1988.      19.5      23.1
 418.       28.6      1988.      19.5      23.1
 430.       28.6      1988.      19.5      23.1
 438.       28.6      1988.      19.5      23.1
 447.       28.6      1988.      19.5      23.1
 448.       28.6      1988.      19.5      23.1
```

Appendix B

ORDER FOR KEY PUNCHING: COMBINED PROGRAMS

Card 1 Program forces and moments, external forces, and graphs.

Card 2 Title card.

Card 3 Number of segments (col. 1).

Card 4 Number of positions (col. 3 and 4).

Card 5 Starting time (col. 1–10) and time interval (col. 11–20).

Card 6 Segment number of the trunk (col. 1), normal trunk length (col. 2–11). Column 12 is left blank if the trunk r is distal, and 1 (one) placed in column 12 if the trunk r is proximal. The number of the segment to be fully key punched (wt., θ, l, r, k) is punched in column 13, and if a second segment is to be key punched, the number of this segment is punched in column 14. The special key punching is used when the weight of a segment changes in the middle of a motion, as when releasing a ball, or if the trunk motion requires the cardboard-lead method for obtaining the trunk center of gravity due to extreme flexion or extension. An alternate method of key punching a released ball is to consider the ball an external force (see card 17).

Card 7 The percentage of the first segment length to obtain r (col. 1–10), the percentage of the first segment length to obtain k (col. 11–20), then alternate r and k percentages for the remaining segments in successive groups of ten columns (40% is key punched .40) (see Table 3-4).

Card 8 Continuation of card 7 if the motion has more than three segments (the percentage to obtain k for segment 4 will be in col. 1–10).

Card 9 Normal segment lengths for each segment (1st segment in col. 1–10, 2nd segment in col. 11–20, etc.).

Card 10 Camera angle correction number taken from the correction tables (col. 1, 2, 3). (see Table B-1 and Fig. B-2.)

Card 11 Weight for each segment (1st segment in col. 1–10, 2nd segment in col. 11–20, etc.).

Card 12 Polynomial for each segment (1st segment in col. 1, 2nd segment in col. 2, 3rd segment in col. 3, etc.).

Card 13 Scale number for plotting the velocity and acceleration curves (1st segment in col. 1–10, 2nd segment in col. 11–20, etc.). The scale number should be just larger than the largest absolute acceleration given in degrees per second.

Card 14 Scale numbers for F_x, F_y, and moments for 1st segment. Minimum F_x in col. 1–8, maximum F_x in col. 9–16, minimum F_y in col. 17–24, maximum F_y in col. 25–32, minimum moments in col. 33–40, maximum moments in col. 41–48, and the segment number is punched in col. 49. The key punching must be in the following form:

$$-1.0 + 004 + 1.0 + 004 - 1.0 + 004 + 1.0 + 004 - 1.0 + 007 + 1.0 + 0071$$

Card 15 Scale numbers for F_x, F_y, and moments for the 2nd segment.

Card 16 Scale numbers for F_x, F_y, and moments for the 3rd segment. Continue on separate cards for each segment.

Card 17 First data card. θ is punched in col. 1–10, and L (length) in col. 11–19 for segment 1, position 1. Column 20 is left blank if the segment motion is in front of the plane, as determined by the placement of the scale in the pictures, and a 1 (one) is placed in col. 20 if the segment motion is behind the plane. External forces on this segment. F_x is punched in col. 51–60, F_y in col. 61–70, and the distance of the force application from the pivot point (joint) in col. 71–80.

If a segment is fully key punched, the angle θ is punched in col. 1–10, segment length (L) in col. 11–19, a 1 (one) or blank in col. 20, the segment weight in col. 21–30, r in col. 31–40, and k in col. 41–50.

Card 18 Second data card for segment 1, position 2.
Card 19 Third data card for segment 1, position 3. Continue on separate cards to complete the number of positions for segment 1.
Card 20 Data card for segment 2, position 1.
Card 21 Data card for segment 2, position 2.
Card 22 Data card for segment 2, position 3. Continue on separate cards to complete the number of positions for segment 2.
Last card Punch appropriate end card.

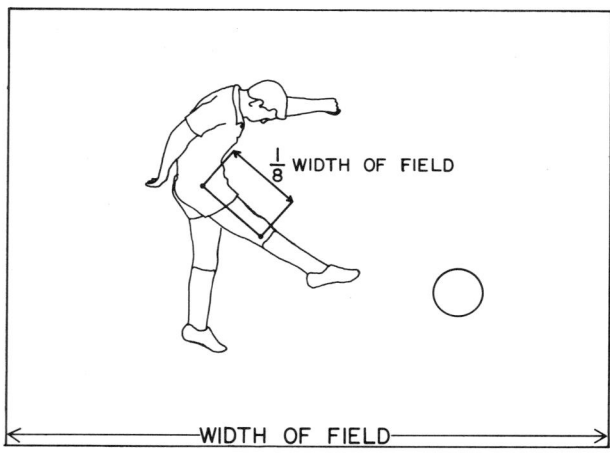

Fig. B-2
Camera angle correction number for computer program.

Table B-1 Number for Camera Angle Correction

Lens and Distance	If segment fills $\frac{1}{4}$ of width of field	If segment fills $\frac{1}{8}$ of width of field
	NUMBER TO BE ENTERED ON CONTROL CARD	
15mm at $18\frac{1}{2}$ ft	154	158
25mm at 30 ft	254	258
50mm at 60 ft	504	508
75mm at 93 ft	754	758

Card 10 chooses the percentage correction in the measured length of a line as determined by the lens used, the camera to subject distance, and the size of the image (Fig. B-2). The first two numbers on the card represent the lens used and the third, the relationship of the body segment measured to the width of the field. Therefore, if 508 is punched, a 50mm lens was used resulting in a picture having the body segment $\frac{1}{8}$ the width of field. The appropriate percentage corrections are programmed according to Table B-1.

Appendix B

COMPUTER OUTPUT

The output (see Table B-2) first lists the absolute angular velocities and accelerations in degrees per second, and graphs the data according to the scale choosen on card 13. This program uses a least-squares polynomial curve fit, taken from Kuo (C-16), to determine the velocities and accelerations. Polynomial degree 4 should be used for smooth, slow motions, while more abrupt changes in motions would require degree 5 to 9. The polynomial used is determined by comparing the input angles with the output angles as punched by the computer. If the input and output angles are not within a few degrees of each other, the polynomial should be increased. If they are within three or four degrees of each other, the calculated angular velocities and accelerations are accepted.

The total body center of gravity is given as an X and Y distance in centimeters from the fixed point of the link system measured.

Five columns of data for each segment and each position are listed as:

F_x = horizontal force in grams.
F_y = vertical force in grams.
F = resultant force of F_x and F_y.
Phi = direction of the resultant force in degrees from the right horizontal.
Moments = joint moments in gram-centimeters.

Next the segment velocities and accelerations are zeroed in successive order to determine the contribution of the motion of each segment to the whole motion. The numbers represent a percentage change in the horizontal force (F_x), vertical force (F_y), and moments. Therefore, the line indicated by the arrow in Table B-2 would read: For position 5, with ω (relative angular velocity) and α (acceleration) of segment 1 zeroed, the change in the F_x for segment 1 is 87.39% less than 20,721 gm, the change in F_y for segment 1 is 38% less than 3442 gm, and the change in the moment of segment 1 is 49.24% less than 21,005 gm-cm. Therefore, the largest percentage change with a minus sign ($-$), indicates the segment that is the largest contributing factor relative to the whole motion.

Table B-2 Computer Output

```
5 CGX= 2.49526092+001  CGY= -7.68659205+000


           REL. VEL.            REL. ACC.
   1         8.63                -4.34
   2        10.12              -105.98
           FX              FY              F              PHI     MOMENT
 11  5  2.07217668+004  3.44279751+003  2.10058200+004   9.43  -5.07210038+005
 12  5  1.85013721+004  7.70821831+003  2.00428890+004  22.62  -3.89360811+004
 SEG, ZEROED   FX       FY       Mo
   5  1   1    -87.39   -38.00   -49.24  <----
   5  1   2    -85.87   -35.99   227.77

   5  2   1    -52.67  -317.35  -120.52
   5  2   2    -58.99  -141.74  -447.28
```

A summary relisting of the forces and moments, and velocities and accelerations, with graphs of the forces and moments, is placed at the end of the computer output.

The following program (Table B-3) was written by Myron Curtis, Bowdoin College, Brunswick, Maine. His work was condensed, modified, and then expanded by Gerald Calkins, University of Massachusetts, Amherst, Massachusetts, to include camera angle corrections and the possibility of applying external forces.

MEANING OF VARIABLES

1. WHOA and WHOB — Used to store title statements. Each is 80 spaces alphanumeric.
2. NSEG — The number of segments being analyzed.
3. NPOS — The number of positions being analyzed.
4. XMIN — The starting time (usually 0.0).
5. DELTX—The amount of time in seconds between positions.
6. NTRK — The number of the trunk segment.
7. TRNKNL — The normal trunk length.
8. KIP — A zero stored in KIP means the trunk is rotating around the shoulder. A one stored in KIP means the trunk is rotating around the hip.
9. NSPEC, NSPEC1 — The number of the segments (up to two) which will have their weight, R, K, punched separately on each data card.
10. PCTR, PCTK — The percent of the segment length which equals R and K.
11. EN — The normal segment length.
12. COR — A number to indicate the lens and distance (for camera angle correction).
13. W — The weight of the segment.
14. MP — The polynomial number.
15. DATY — The angular position of the segment.
16. DATL — The length of the segment.
17. NFBD — A zero indicates motion is in front of the plane; a one indicates the motion is behind the plane (for camera angle correction).
18. DATW — The weight of the segment.
19. DATR — The distance to the center of gravity of the segment.
20. DATK — The radius of gyration of the segment.
21. FXE, FYE — The x and y components of the external force.
22. RE — The distance from the point of application of the external force to the joint.

Table B-3 Computer Program

```
C      (1) COMPUTE THE M + 1 VALUES OF XBAR (I), WHERE M IS THE DEGREE
C          OF THE POLYNOMIAL Y(M).
C      (2) NORMALIZE THE INITIAL VALUES OF X(I) TO THE INTERVAL (-1,1).
C      (3) PERFORM THE LAGRANGIAN INTERPOLATION TO OBTAIN M+ 1 VALUES OF
C          YBAR(I) WHICH CORRESPOND TO THE M + 1 VALUES OF THE XBAR(I).
C      (4) COMPUTE THE COEFFICIENTS C(I).
C      (5) CONVERT THE CHEBYSHEV SERIES FOR Y(M) TO ITS EQUIVALENT POWER
C          SERIES.
C      (6) CONVERT THE POWER SERIES FROM THE INTERVAL (-1,1) TO THE
C          INTERVAL (A,B).
C      (7) PUNCH THE COEFFICIENTS OF THE FINAL SERIES EXPANSION.
C      M = DEGREE OF THE POLYNOMIAL Y(M) DESIRED.
C      XMIN = FIRST VALUE OF X (SMALLEST VALUE OF ORIGINAL X-COORDINATES)
C      DELTX = INCREMENT BETWEEN VALUES OF X, THAT IS, (X(I) - X(I) - 1).
C      Y(J) = VALUE OF THE ORIGINAL Y CORRESPONDING THE JTH VALUE OF X.
C      R(I) = THE ITH ROOT, OR XBAR(I).
C      V(I) = THE ITH VALUE OF XP(I), OR NORMALIZED X(I).
C      C(I) = THE ITH COEFFICIENT OF THE CHEBYSHEV SERIES IN (-1,1).
C      F(I) = THE INTERMEDIATE STORAGE USED IN COMPUTING INTERPOLATED
C      YBAR(I), IN COMPUTING C(I)@S, AND IN CONVERTING C(I)@S TO FINAL
C      POWER-SERIES COEFFICIENTS IN (A,B).  THE FINAL COEFFICIENTS ARE
C      STORED IN Y(J).
C      CHEBYSHEV POLYNOMIAL APPROXIMATION - EQUIDISTANT DATA
       DIMENSION Y1(90),DATTH(8,50),DATV(8,50),DATA(8,50),NFBD(8,50)
       DIMENSION S(20),V(90),Y(90),C(20),F(20),DATY(8,50),DATL(8,50)
       DIMENSION YGRAPH(4),IC(4),DATW(8,50),DATR(8,50),DATK(8,50)
       DIMENSION W(8),XL(8),R(8),A(8),B(8),XMASS(8),CG(8,2),Z(8,2)
       DIMENSION PCTR(8),PCTK(8), EN(8),NFB(8),CIS(8),CXL(8),DATM(8,50)
       DIMENSION DUMW(8),DUMR(8),DUMK(8),WHOA(10),WHOB(10),MP(8),YMAXX(8)
      1,OMEGA(8),ALPHA(8),OMEG(8),ALPH(8),FX(8),FY(8),XMOMT(8)
      2,FXA(8),FYA(8),AMOMT(8),XK(8),IZ(8),DFX(8,50)
      3,FXE(8,50),FYE(8,50),  XFI(8),XFA(8),YFI(8),YFA(8),MI(8),MA(8)
      4,DFY(8,50),RE(8,50),RR(8,8),AA(8,8),THETA(8),STORE(5,50,8)
       COMMON PI,CONST,W,XL,XK,R,A,B,XMASS,CG,Z,OMEGA,ALPHA,OMEG,ALPH
      1,NSEG,IT,FXE,FYE,NPOS,RE,RR,AA,THETA
       EQUIVALENCE(YGRAPH(1),X1),(YGRAPH(2),X2),(YGRAPH(3),X3)
     1 READ 300,WHOA
       IF(EOF,60)9999,9998
  9998 READ 300,WHOB
   300 FORMAT(10A8)
       PRINT 301,WHOA,WHOB
   301 FORMAT (///1X,10A8/1X,10A8)
       PRINT 302
   302 FORMAT(* ANG.= DEG., VEL.= DEG. PER SEC., ACC.= DEG. PER SEC. SQ.*
      1)
       READ 5,NSEG, NPOS,XMIN,DELTX
     5 FORMAT(I1/I4/2F10.5)
       READ 104,NTRK,TRNKNL,KIP,NSPEC,NSPEC1
   104 FORMAT (I1,F10.3,3I1)
       READ 101,(PCTR(I),PCTK(I),I=1,NSEG)
       READ 101,( EN(I),I=1,NSEG)
   101 FORMAT(7F10.3)
       READ 136,COR
   136 FORMAT(I3)
       READ 101,(W(I),I=1,NSEG)
       READ 303,(MP(ID),ID=1,NSEG)
   303 FORMAT(7I1)
       READ 101,(YMAXX(ID),ID=1,NSEG)
       DO 3000 I=1,NSEG
  3000 READ 3010,XFI(I),XFA(I),YFI(I),YFA(I),MI(I),MA(I),IZ(I)
  3010 FORMAT(6E8.1,A2)
```

```
            NET=1
            N=NPOS
            IND=NSEG+1
            DO 700 ID=1,NSEG
            DO 700 JD=1,NPOS
        700 READ 710,DATY(ID,JD),DATL(ID,JD),NFBD(ID,JD),DATW(ID,JD),DATR(ID,J
           1D),DATK(ID,JD),FXE(ID,JD),FYE(ID,JD),RE(ID,JD)
        710 FORMAT(F10.5,F9.5,I1,6F10.5)
            ID=1
        138 M=MP(ID)
            YMAX=YMAXX(ID)
            YMIN=-YMAX
            XMIN=0.0
            XMIN1=XMIN
            M = M + 1
C           COMPUTE ROOTS
            X2 = M
            DO 1000 I=1,M
            X1 = I
            ARG=1.5707963268*(2.0*X1-1.0)/X2
            MSUB = M+1-I
            S(MSUB) = COSF(ARG)
       1000 CONTINUE
C           NORMALIZE VECTORS
            X1=N-1
            DV=2.0/X1
            V(1)= -1.0
            L = N-1
            DO 1003 I=1,L
            V(I+1) = V(I)+DV
       1003 CONTINUE
            DO 720 JD=1,NPOS
            J=JD
        720 Y(J)=DATY(ID,JD)
            DO 11 I23=1,N
         11 Y1(I23)=Y(I23)
C           PERFORM LAGRANGIAN INTERPOLATION
            I=1
            DO 150 L=1,N
            IF(S(I)-V(L)) 151,151,150
        151 U=(S(I)-V(L-1))/(V(L)-V(L-1))
            IF (L-2)154,154,155
        154 F(I)=U*(Y(L)-Y(L-1)) + Y(L-1)
            GO TO 157
        155 IF (L-N) 156,154,154
        156 ZIP = -U*(U-1.)*(U-2.)*Y(L-2)/6. + (U*U-1.)*(U-2.)*Y(L-1)/2.
            F(I) = ZIP - (U+1.)*(U-2.)*U*Y(L)/2. + U*(U*U-1.)*Y(L+1)/6.
        157 I=I+1
            IF (M+1-I)153,153,160
        160 IF (S(I)-V(L))151,151,150
        150 CONTINUE
C           COMPUTE COEFFICIENTS
        153 ZM = M
            DO 280 I=1,M
            SUM =0.0
            IF (I-2) 260,265,270
        260 DO 261 J=1,M
        261 SUM = SUM + F(J)
            GO TO 275
        265 DO 266 J=1,M
        266 SUM = SUM + S(J)*F(J)
```

```
        GO TO 275
270     V(1)=1.0
        DO 272 J=1,M
        V(2)=S(J)
        DO 271 K=3,I
271     V(K)=2.*S(J)*V(K-1) - V(K-2)
272     SUM = SUM + F(J)*V(I)
275     C(I) = 2.0*SUM/ZM
280     CONTINUE
        C(1) = C(1)/2.0
        ZIP = XMIN
        NN = N-1
        DO 593 J=1,NN
C
593     XMIN = XMIN - DELTX
        SUM1=(XMIN+ZIP)/2.
        SUM2=(XMIN-ZIP)/2.
        NN = M-1
        PRINT 305, NN
305     FORMAT (33HPOLYMONIAL COEFICIENTS - DEGREE ,I2,//)
C       CONVERT CHEBYSHEV SERIES TO IT@S EQUIVALENT POWER SERIES
        F(1)=C(1)
        F(2)=C(2)
        IF (M-2)597,597,596
596     DO 598 K=1,M
        V(K)=0.0
        Y(K)=0.0
598     F(K+2)=0.0
        V(2)=1.0
        DO 599 K=3,M
        ZIP=K-1
        Y(1)=COSF(3.14159265*ZIP/2.)
        DO 594 J=2,K
594     Y(J)=2.*V(J-1) - Y(J)
        DO 595 J=1,K
        F(J) = F(J) - C(K)*Y(J)
        ZIP=V(J)
        V(J) = Y(J)
595     Y(J) = ZIP
599     CONTINUE
C       GO BACK TO ORIGINAL INTERVAL
597     Y(1) = F(1)
        DO 580 K=2,M
580     Y(K)=0.0
        DO 583 K=2,M
        L=K-1
        Y(K) = Y(K) + F(K)/SUM2**L
        SUM3=1.0
        ZIP=1.0
        X2=K
        DO 582 J=1,L
        X1=J
        SUM3=SUM3*X1
        DV=SUM3*SUM2**L
        ZIP=ZIP*(X2-X1)
        KMJ=K-J
582     Y(KMJ)=Y(KMJ) + (ZIP*SUM1**J*(-1.)**J*F(K))/DV
583     CONTINUE
C       COEFFICIENTS STORED IN Y(K)
        DO 584 K=1,M
        I=K-1
```

```
      584  PRINT 585,I,Y(K)
      585  FORMAT (30X,2HA(,I2,3H) =,E16.8)
           DO 2 I=1,M
           J=M-I
        2  C(I)=Y(J+1)
           CALL DERIV1(M-1,C,V)
           CALL DERIV1(M-2,V,F)
           X=XMIN1
           SUM=0.0
           IC(1)=1HT
           IC(2)=1HV
           IC(3)=1HA
           DO 4 I=1,NPOS
           JD=I
           X1=SYND1(M-1,C,S,X)
           PRINT 501,Y1(I),X1
      501  FORMAT(1X,2F10.5)
           SUM=SUM+(Y1(I)-X1)**2
       40  X2=SYND1(M-2,V,S,X)
           X3=SYND1(M-3,F,S,X)
      500  FORMAT(10X,F10.3,30X,2F10.1,I2,I8)
           CALL PLOT1(X,YGRAPH,IC,3,YMIN,YMAX,I)
           PRINT 5678,X2,X3
     5678  FORMAT(100X,2F10.1)
           DATTH(ID,JD)=X1
           DATV(ID,JD)=X2
           DATA(ID,JD)=X3
        4  X=X+DELTX
           SUM=SQRTF(SUM)
           PRINT 502,SUM
      502  FORMAT(1H0,15HERROR IN FIT IS,E16.8)
           NFT=NFT+1
           ID=ID+1
           IF(NFT.EQ.IND)GO TO 1112
           GO TO 138
     1112  CONTINUE
    C      PROGRAM TWO
    C      BODY MOTION ANALYSIS RELATIVE AND ABSOLUTE METHOD    STAN PLAGENHOEF
    C      INPUT NUMBER OF SEGMENTS FORMAT( I 1)
    C      WEIGHT,THETA,X,R,K,OMEGA,ALPHA    WHERE OMEGA AND ALPHA ARE ABSOLUTE
    C      THESE QUANTITIES A
     2005  Z(1,1)=0.0
           Z(1,2)=0.0
           PI = 3.1415927
           CONST = .017453293
           PRINT 304
      304  FORMAT(///* VEL.= RAD. PER SEC., ACC.= RAD. PER SEC. SQ.*/* FX,FY,
          1 AND F = GM., PHI = DEG., MOMENT = GM.-CM.*/* CGX AND CGY = CM.*)
           N1 = NSEG-1
           IT=1
           NPOINT=NPOS
           IF(COR.EQ.154)GO TO 815
           IF(COR.EQ.158)GO TO 816
           IF(COR.EQ.254)GO TO 825
           IF(COR.EQ.258)GO TO 826
           IF(COR.EQ.504)GO TO 850
           IF(COR.EQ.508)GO TO 851
           IF(COR.EQ.754)GO TO 875
           IF(COR.EQ.758)GO TO 876
      815  AC15=.056
           AC30=.111
```

```
      AC45=.5
      AC60=.97
      AC75=.226
      GO TO 880
  816 AC15=.028
      AC30=.055
      AC45=.075
      AC60=.098
      AC75=.113
      GO TO 880
  825 AC15=.032
      AC30=.068
      AC45=.091
      AC60=.122
      AC75=.142
      GO TO 880
  826 AC15=.016
      AC30=.034
      AC45=.045
      AC60=.061
      AC75=.071
      GO TO 880
  850 AC15=.021
      AC30=.036
      AC45=.047
      AC60=.074
      AC75=.083
      GO TO 880
  851 AC15=.01
      AC30=.018
      AC45=.023
      AC60=.037
      AC75=.041
      GO TO 880
  875 AC15=.0093
      AC30=.02
      AC45=.024
      AC60=.04
      AC75=.045
      GO TO 880
  876 AC15=.0046
      AC30=.01
      AC45=.012
      AC60=.02
      AC75=.022
  880 CONTINUE
      DO 2014 I2=1,NPOS
      JD=I2
 2007 CGX=0.0
      CGY=0.0
      SW=0.0
      DO 2011 I=1,NSEG
      ID=I
      NFB(I)=NFBD(ID,JD)
      OM=DATV(ID,JD)
      ALP=DATA(ID,JD)
      DUMW(I)=DATW(ID,JD)
      XL(I)=DATL(ID,JD)
      DUMR(I)=DATR(ID,JD)
      DUMK(I)=DATK(ID,JD)
      IF(I.EQ.NSPEC)GO TO2400
```

```
            IF(I.EQ.NSPEC1)GO TO 2400
            IF (I.EQ.NTRK) GO TO 2019
            CXL(I)=ABS(XL(I)-EN(I))
            IF(CXL(I).LT.1.0)GO TO 2020
            CIS(I)=XL(I)/ EN(I)
            IF(CIS(I).LT..97)GO TO2303
            IF(CIS(I).LT..87)GO TO2305
            IF(CIS(I).LT..71)GO TO2307
            IF(CIS(I).LT..5 )GO TO2309
            IF(CIS(I).LT..26)GO TO2311
       2303 IF(NFB(I).EQ.1)GO TO2304
            XL(I)=XL(I)-XL(I)*AC15
            GO TO2302
       2304 XL(I)=XL(I)+XL(I)*AC15
            GO TO2302
       2305 IF(NFB(I).EQ.1)GO TO2306
            XL(I)=XL(I)-XL(I)*AC30
            GO TO2302
       2306 XL(I)=XL(I)+XL(I)*AC30
            GO TO2302
       2307 IF(NFB(I).EQ.1)GO TO2308
            XL(I)=XL(I)-XL(I)*AC45
            GO TO2302
       2308 XL(I)=XL(I)+XL(I)*AC45
            GO TO2302
       2309 IF(NFB(I).EQ.1)GO TO2310
            XL(I)=XL(I)-XL(I)*AC60
            GO TO2302
       2310 XL(I)=XL(I)+XL(I)*AC60
            GO TO2302
       2311 IF(NFB(I).EQ.1)GO TO2312
            XL(I)=XL(I)-XL(I)*AC75
            GO TO2302
       2312 XL(I)=XL(I)+XL(I)*AC75
       2302 IF(I-NTRK) 2020,2019,2020
       2019 IF(XL(I)-TRNKNL)2020,2020,2021
       2021 IF(KIP.EQ.1)GO TO2121
            R(I)=TRNKNL*PCTR(I)+((XL(I)-TRNKNL)-(XL(I)-TRNKNL)*0.018
            XK(I)=TRNKNL*PCTK(I)+((XL(I)-TRNKNL)-(XL(I)-TRNKNL)*0.018)
            GO TO 2022
       2121 R(I)=TRNKNL*PCTR(I)+(XL(I)-TRNKNL)*0.018
            XK(I)=TRNKNL*PCTK(I)+(XL(I)-TRNKNL)*0.018
            GO TO 2022
       2020 R(I)=XL(I)*PCTR(I)
            XK(I)=XL(I)*PCTK(I)
            GO TO 2022
       2400 W(I)=DUMW(I)
            R(I)=DUMR(I)
            XK(I)=DUMK(I)
       2022 CONTINUE
            XMASS(I)=W(I)/980.0
            THET =DATTH(ID,JD)
            THETA(I)=THET*CONST
            OMEGA(I) = OM*CONST
            ALPHA(I) = ALP*CONST
            A(I) = SIN(THETA(I))
            B(I) = COS(THETA(I))
            Z(I+1,1)=Z(I,1) + XL(I)*B(I)
            Z(I+1,2) = Z(I,2) + XL(I)*A(I)
            CG(I,1) = Z(I,1) + R(I)*B(I)
            CG(I,2) = Z(I,2) + R(I)*A(I)
```

```
              DO 5555J=1,NSEG
              RR(J,I)=SQRT(((CG(I,1)-Z(J,1))**2)+(CG(I,2)-Z(J,2))**2)
       5555   AA(J,I)=ATAN((CG(I,2)-Z(J,2))/(CG(I,1)-Z(J,1)))
              CGX=CGX+W(I)*CG(I,1)
              CGY=CGY+W(I)*CG(I,2)
       2011   SW=SW+W(I)
              CGX=CGX/SW
              CGY=CGY/SW
              PRINT 2201,IT,CGX,CGY
              OMEG(1)=OMEGA(1)
              ALPH(1)=ALPHA(1)
              PRINT 2800
       2800   FORMAT(//10X,*REL. VEL.*10X,*REL. ACC.*)
              K=1
              PRINT 2900,K,OMEG(1),ALPH(1)
              DO 2012 I=2,NSEG
              OMEG(I)=OMEGA(I)-OMEGA(I-1)
              ALPH(I)=ALPHA(I)-ALPHA(I-1)
       2012   PRINT 2900,I,OMEG(I),ALPH(I)
              DO 410 I=1,NSEG
              STORE (2,IT,I)=ALPH(I)
       410    STORE (5,IT,I)=OMEG(I)
       2900   FORMAT(3X,I1,2X,F10.2,9X,F10.2)
              CALL MOTION(1,FXA,FYA,AMOMT,I2)
              DO 420 I=1,NSEG
              STORE (1,IT,I)=AMOMT(I)
              STORE (3,IT,I)=FYA(I)
       420    STORE (4,IT,I)=FXA(I)
              DO 7001 I=1,NSEG
              DFX(I,I2)=FXA(I)
              DFY(I,I2)=FYA(I)
       7001   DATM(I,I2)=AMOMT(I)
              DO 2013 IJ=1,NSEG
              SAVEO=OMEG(IJ)
              SAVEA=ALPH(IJ)
              ALPH(IJ)=0.
              OMEG(IJ)=0.
              CALL MOTION(0,FX,FY,XMOMT,I2)
              OMEG(IJ)=SAVEO
              ALPH(IJ)=SAVEA
              PRINT 2102
              DO 2013 IJ1=1,NSEG
              PCTX=(FX(IJ1)/FXA(IJ1)-1.0)*100.0
              PCTY=(FY(IJ1)/FYA(IJ1)-1.0)*100.0
              PCTM=(XMOMT(IJ1)/AMOMT(IJ1)-1.0)*100.0
       2013   PRINT 2200,IT,IJ,IJ1,PCTX,PCTY,PCTM
              IT=IT+1
       2014   CONTINUE
       499    PRINT 5001
       5001   FORMAT (//30X,8H MOMENTS/)
              PRINT 5011,(I,I=1,NSEG)
       5011   FORMAT (5H POS.,6X,8(I2,4H SEG,10X))
              DO 504 IT=1,NPOINT
       504    PRINT 505, IT,(STORE(1,IT,I), I=1,NSEG)
       505    FORMAT(1X,I2,4X,8E16.8)
              PRINT 506
       506    FORMAT (//30X,22H RELATIVE ACCELERATION/)
              PRINT 5011,(I,I=1,NSEG)
              DO 508 IT=1,NPOINT
       508    PRINT 505, IT,(STORE(2,IT,I), I=1,NSEG)
              PRINT 510
```

```
510   FORMAT (//30X,19H FY, VERTICAL FORCE/)
      PRINT 5011,(I,I=1,NSEG)
      DO 512 IT=1,NPOINT
512   PRINT 505, IT, (STORE(3,IT,I), I=1,NSEG)
      PRINT 515
515   FORMAT (//30X,21H FX, HORIZONTAL FORCE/)
      PRINT 5011,(I,I=1,NSEG)
      DO 517 IT=1,NPOINT
517   PRINT 505, IT, (STORE(4,IT,I), I=1,NSEG)
      PRINT 519
519   FORMAT (//30X,18H RELATIVE VELOCITY/)
      PRINT 5011,(I,I=1,NSEG)
      DO 520 IT=1,NPOINT
520   PRINT 505, IT,(STORE(5,IT,I), I=1,NSEG)
      CALL PLOT (NSEG,NPOS,0,DELTX,XFI,XFA,IZ,DFX)
      CALL PLOT (NSEG,NPOS,0,DELTX,YFI,YFA,IZ,DFY)
      CALL PLOT (NSEG,NPOS,0,DELTX,MI,MA,IZ,DATM)
      GO TO 1
2100  FORMAT(I1)
2102  FORMAT(1H )
2200  FORMAT(3X,I3,2I2,3F9.2)
2201  FORMAT(1H0,2X,I3,5H CGX=E16.8,5H CGY=E16.8)
9999  STOP
      END
      FUNCTION SYND1(N,A,B,X)
      DIMENSION A(N),B(N)
      B(1)=A(1)
      IF(N-1)2,2,30
30    DO 1 I=2,N
1     B(I)=A(I)+X*B(I-1)
3     SYND1=A(N+1)+X*B(N)
      RETURN
2     SYND1=B(1)
      RETURN
      END
      SUBROUTINE DERIV1(N,A,B)
      DIMENSION A(N),B(N)
      IF(N-1)3,3,2
2     DO 1 I=1,N
      XN=N+1-I
1     B(I)=A(I)*XN
      RETURN
3     B(1)=0.0
      RETURN
      END
      SUBROUTINE PLOT1(X,Y,IC,N,YMIN,YMAX,IT)
C PLOTS X,Y(I),IC(I) IS ITH CHARAXTERTO PLOT
      DIMENSION Y(10),IC(10),LINE(110)
      DY=(YMAX-YMIN)/109.
      DO 2 J=1,110
2     LINE(J)=1H
      LINE(55)=7000
      DO 1 I=1,N
      J=(Y(I)-YMIN)/DY
      J=J+1
      IF(J)3,3,4
4     IF(J-109)6,6,5
6     CONTINUE
1     LINE(J)=IC(I)
      PRINT 100,X,IT,(LINE(I),I=1,110)
100   FORMAT(1X,F6.3,I3,99A1,11A1)
```

```
            RETURN
    3       J=1
            GO TO 6
    5       J=110
            GO TO 6
            END
            SUBROUTINE MOTION (L,FX,FY,XMOMT,I2)
            DIMENSION W(8),XL(8),XK(8),R(8),A(8),B(8),XMASS(8),CG(8,2),Z(8,2)
           1,OMEGA(8),ALPHA(8),OMEG(8),ALPH(8),FX(8),FY(8),XMOMT(8)
           2,FXE(8,50),FYE(8,50),RE(8,50),RR(8,8),AA(8,8),THETA(8),S9(8)
            COMMON PI,CONST,W,XL,XK,R,A,B,XMASS,CG,Z,OMEGA,ALPHA,OMEG,ALPH
           1,NSEG,IT,FXE,FYE,NPOS,RE,RR,AA,THETA
            FX(NSEG+1) = 0.0
            FY(NSEG+1) = 0.0
            XMOMT(NSEG+1) = 0.0
            I = NSEG
            DO 5 J=1,NSEG
            JD=I2
            ID=I
            FY(I)=FORCE(I,1,L,0)+FY(I+1)+FYE(ID,JD)
            FX(I)=FORCE(I,0,L,0)+FX(I+1)+FXE(ID,JD)
            IF(OMEG (I))70,7,7
   70       ALPH(I)=-ALPH(I)
    7       XMOMT(I)=W(I)*R(I)*B(I)+XMASS(I)*ALPH(I)*(XK(I)**2)        +FX(I+1)*
           1XL(I)*A(I)-FY(I+1)*XL(I)*B(I)+XMOMT(I+1)+RE(ID,JD)*(FYE(ID,JD)*B(I)
           2)-FXE(ID,JD)*A(I))
            IF(OMEG(I))72,73,73
   72       ALPH(I)=-ALPH(I)
   73       IF(I.EQ.1)GO TO 5
   22       K=I-1
   23       IF(OMEG (K))80,82,82
   80       ALPH(K)=-ALPH(K)
   82       CC=COSF(THETA(I)-AA(K,I))
            SS=SINF(THETA(I)-AA(K,I))
            XMOMT(I)=XMOMT(I)+XMASS(I)*RR(K,I)*R(I)*((OMEG(K)**2)*SS-ALPH(K)*C
           1C)
            IF(OMEG(K))83,84,84
   83       ALPH(K)=-ALPH(K)
   84       IF(K.EQ.1)GO TO 5
            K=K-1
            GO TO 23
    5       I = I-1
            IF(L) 9,17,9
    9       PRINT 103
            DO 15 I=1,NSEG
            IF(FX(I)) 10,11,10
   10       PHI = ATAN(FY(I)/FX(I))
            IF(FX(I)) 12,13,13
   11       PHI = 90.
            GO TO 14
   12       PHI = PHI+PI
   13       PHI = PHI/CONST
   14       F = SQRT(FX(I)**2+FY(I)**2)
   15       PRINT 102,L,I,IT,FX(I),FY(I),F,PHI,XMOMT(I)
   17       RETURN
  102       FORMAT(1X,I1,I1,I3,3E16.8,F9.2,E16.8)
  103       FORMAT(1H0,11X,2HFX,14X,2HFY,15X,1HF,13X,3HPHI,8X,6HMOMENT)
            END
            FUNCTION FORCE (J,K,L,L1)
            DIMENSION W(8),XL(8),XK(8),R(8),A(8),B(8),XMASS(8),CG(8,2),Z(8,2)
           1,OMEGA(8),ALFHA(8),OMEG(8),ALPH(8),S10(8)
```

```
      COMMON PI,CONST,W,XL,XK,R,A,B,XMASS,CG,Z,OMEGA,ALPHA,OMEG,ALPH
      IF(K.EQ.0)GO TO 2
      TR=A(J)
      GO TO 3
    2 TR=B(J)
    3 CONTINUE
      J1=J-1
      SGN = (-1)**K
      CGK1 = CG(J,K+1)
      K2 = 2-K
      CGK2 = CG(J,K2)
   20 FORCE = 0.0
      DO 51 I=1,J
      IF(OMEG (I))10,21,21
   10 ALPH(I)=-ALPH(I)
   21 FORCE =FORCE+OMEG(I)**2*(CGK1-Z(I,K+1))+SGN*ALPH(I)*(CGK2-Z(I,K2))
      IF(OMEG(I))50,51,51
   50 ALPH(I)=-ALPH(I)
   51 CONTINUE
      IF(J1) 22,30,22
   22 DO 24 I=1,J1
      SUM1=0.0
      DO 23 K1=I,J1
      IND=0
      SIGE=OMEG(J)*OMEG(I)
      IF(SIGE)200,201,201
  200 TR=-TR
      IND=1
  201 SUM1=SUM1+ABSF(OMEG(I))*ABSF(OMEG(K1+1))*R(K1+1)*TR
      IF(IND.EQ.0)GO TO 23
      TR=-TR
   23 CONTINUE
  100 IF(J.LT.3)GO TO 24
      SUM2=0.
      J14=J-2
      DO 110 K14=1,J14
      K16=K14+1
      J15=J-1
      DO 110 K15=K16,J15
      INDE=0
      SIGN=OMEG(K14)*OMEG(J)
      IF(SIGN)101,102,102
  101 TR=-TR
      INDE=1
  102 OM14=ABSF(OMEG(K14))
      OM15=ABSF(OMEG(K15))
      SUM2=SUM2+OM14*OM15*R(J)*TR
      IF(INDE.EQ.0)GO TO 110
      TR=-TR
  110 CONTINUE
      J4=J
  115 IF(J4.LT.4)GO TO 124
      J40=J4-3
      DO 120 K4=1,J40
      K6=K4+1
      J5=J40+1
      DO 120 K5=K6,J5
      INDA=0
      SIGA=OMEG(K4)*OMEG(J)
      IF(SIGA)116,117,117
  116 TR=-TR
```

```
          INDA=1
      117 OM4=ABSF(OMEG(K4))
          OM5=ABSF(OMEG(K5))
          SUM2=SUM2+OM4*OM5*XL(J4-1)*TR
          IF(INDA.EQ.0)GO TO 120
          TR=-TR
      120 CONTINUE
          J4=J4-1
          GO TO 115
      124 FORCE=FORCE+2.*SUM2
       24 FORCE = FORCE+2.*SUM1
       30 FORCE = FORCE*XMASS(J)
          IF(L1)40,41,40
       40 RETURN
       41 IF(K) 31,32,31
       31 FORCE = FORCE-W(J)
       32 RETURN
          END
          SUBROUTINE PLOT(NSEG,NP,X0,DX,YMIN,YMAX,IC,A)
          DIMENSION Y(10),IC(10),YMIN(10),YMAX(10),A(8,100),YINT(10)
          X=X0
          DO 7 I=1,NSEG
    7     YINT(I)=0.0
          PRINT 201,(I,YMIN(I),I,YMAX(I),I=1,NSEG)
      201 FORMAT(* YMIN (*I1,*) =S*E16.8,*  YMAX (*I1,*) =S*E16.8)
          DO 2 J=1,NP
          DO 1 I=1,NSEG
          ID=I
          IT=J
          JD=J
    1     Y(I)=A(ID,JD)
          DO 5 I=1,NSEG
          ODD=NP-(NP/2)*2+1
    5     YINT(I)=2.0*ODD*Y(I)+YINT(I)
          L=7000
          PRINT 300,L,_
      300 FORMAT(64X,A1/64X,A1)
          CALL PLOT2(X,Y,IC,NSEG,YMIN,YMAX,IT)
          X=X+DX
    2     CONTINUE
          DO 6 I=1,NSEG
    6     YINT(I)=YINT(I)*DX/3.0
          PRINT 305,(YINT(I),I=1,NSEG)
      305   FORMAT(26H INTEGRALS OF SEGMENTS ARE,5E16.8)
     9999 RETURN
          END
          SUBROUTINE PLOT2(X,Y,IC,N,YMIN,YMAX,IT)
    C PLOT S X,Y(I),IC(I) IS ITH CHARAXTERTO PLOT
          DIMENSION Y(10),IC(10),LINE(110),YMAX(10),YMIN(10),DY(10)
          DO 8 I=1,N
    8     DY(I)=(YMAX(I)-YMIN(I))/109.
          DO 2 J=1,110
    2     LINE(J)=1H
          LINE(55)=7000
          DO 1 I=1,N
          J=(Y(I)-YMIN(I))/DY(I)
          J=J+1
          IF(J)3,3,4
    4     IF(J-109)6,6,5
    6     CONTINUE
    1     LINE(J)=IC(I)
```

```
      PRINT 100,X,IT,(LINE(I),I=1,110)
100   FORMAT(1X,F6.3,I3   ,99A1,11A1)
      RETURN
3     J=1
      GO TO 6
5     J=110
      GO TO 6
      END
         SCOPE
@LOAD
@RUN,15,15000
```

appendix C

HAND CALCULATION FOR POSITION 5

A hand analysis for position 5 of the basketball throw (Fig. B-1) follows. Both the absolute and relative motions are shown for comparison, and the answers are compared with the computer output. Any differences are due to the slide rule and scaled measurements needed to determine the length of R (the radius from the fixed point of segment 1 to the center of gravity of segment 2; see Fig. C-1). Figures C-2 and C-3 show the free body diagrams for, respectively, the forearm plus hand and the upper arm.

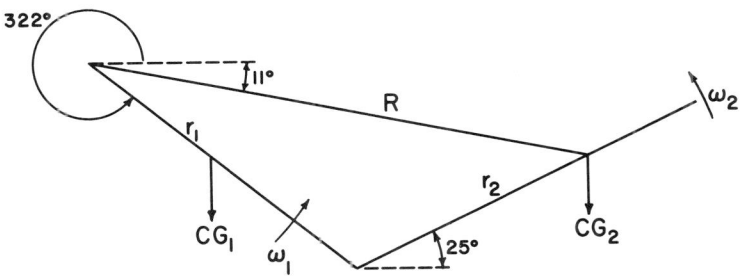

Fig. C-1
Basketball throw — position 5.

Appendix C

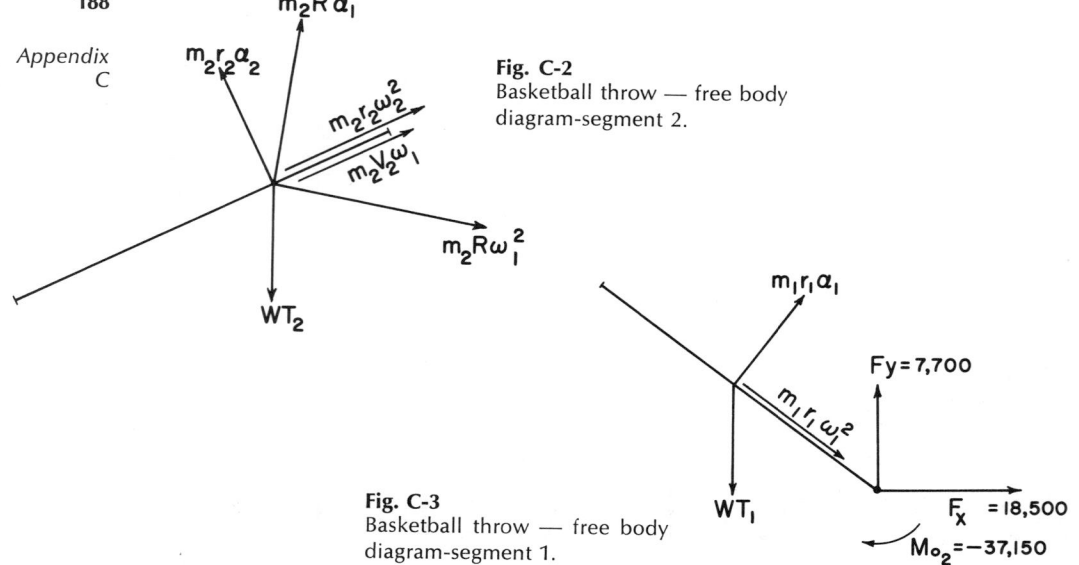

Fig. C-2
Basketball throw — free body diagram-segment 2.

Fig. C-3
Basketball throw — free body diagram-segment 1.

CALCULATIONS FOR POSITION 5 OF BASKETBALL THROW (FIG. B-1)

$WT_1 = 2800 \quad m = 2.858$ $\qquad WT_2 = 2276 \quad m = 2.32$

$\theta_1 = 322°$ $\qquad\qquad\qquad\qquad \theta_2 = 25°$

$\sin 38° = -.615$ $\qquad\qquad \sin 25° = .42$
$\cos 38° = +.79$ $\qquad\qquad \cos 25° = .9$

$\omega_1 \text{abs.} = 8.63$ $\qquad\qquad\quad \omega_2 \text{abs.} = 18.8$

$\omega_1 \text{rel.} = 8.63$ $\qquad\qquad\quad \omega_2 \text{rel.} = 10.12$

$\alpha_1 \text{abs.} = -4.35$ $\qquad\qquad \alpha_2 \text{abs.} = -110$

$\alpha_1 \text{rel.} = -4.35$ $\qquad\qquad \alpha_2 \text{rel.} = -105.98$

$1_1 = 28.6$

$r_1 = 12.5$ $\qquad\qquad\qquad\quad r_2 = 22.4$

$k_1 = 15.5$ $\qquad\qquad\qquad\quad k_2 = 26.6$

$\qquad\qquad R = 43.4 \quad V_2 = r\omega_2$

$\qquad\qquad \phi = 349° \quad V_2 = 22.4 \times 10.12$

$\qquad\qquad\qquad\qquad\quad V_2 = 227$

$\qquad\qquad \sin 11° = .19 \quad \sin 36° = .59$
$\qquad\qquad \cos 11° = .98 \quad \cos 36° = .81$

A free body diagram is drawn to show correct direction of force; therefore do not apply the minus sign of acceleration ($-\alpha$) when solving equations using the relative motion method.

Appendix C

Segment 2 (Absolute)

$$A_{2y} = r_2(\cos\theta_2\alpha_2 - \sin\theta_2\omega_2^2) + l_1(\cos\theta_1\alpha_1 - \sin\theta_1\omega_1^2)$$

$$= 22.4[.9(-110) - .42(18.8)^2] + 28.6[.79(-4.35) - (-.615)(8.63)^2]$$

$$= 22.4(-99 - 149.5) + 28.6(-3.43 + 45.7)$$

$$= 22.4(-248.5) + 28.6(42.3)$$

$$= -5580 + 1210$$

$$A_{2y} = -4370$$

$$m = 2.32 \quad \therefore \quad mA_{2y} = 10,100$$

$$A_{2x} = -r_2(\sin\theta_2\alpha_2 + \cos\theta_2\omega_2^2) - l_1(\sin\theta_1\alpha_1 + \cos\theta_1\omega_1^2)$$

$$= -22.4[.42(-110) + .9(18.8)^2] - 28.6[(-.615)(-4.35) + .79(8.63)^2]$$

$$= -22.4(-46 + 320) - 28.6(2.67 + 58.5)$$

$$= -22.4(274) - 28.6(61.17)$$

$$= -6150 - 1750$$

$$A_{2x} = -7900$$

$$m = 2.32 \quad \therefore \quad mA_{2x} = -18,300$$

Segment 1 (Absolute)

$$A_{1y} = r_1(\cos\theta_1\alpha_1 - \sin\theta_1\omega_1^2)$$

$$= 12.5[.79(-4.35) - (-.615)(8.63)^2]$$

$$= 12.5(-3.44 - 45.7)$$

$$A_{1y} = 530$$

$$m = 2.858 \quad \therefore \quad mA_{1y} = 1515$$

$$A_{1x} = -r_1(\sin\theta_1\alpha_1 + \cos\theta_1\omega_1^2)$$

$$= 12.5[-.615(-4.35) + .79(8.63)^2]$$

$$= -12.5(2.67 + 59)$$

$$= -12.5(61.67)$$

$$A_{1x} = -772$$

$$m = 2.858 \quad \therefore \quad mA_{1x} = -2200$$

Segment 2 (Absolute)

$$F_{2y} = -W_2 - mA_{2y} \qquad F_{2x} = -mA_{2x}$$

$$= -2276 - (-10,100) \qquad = -(-18,300)$$

$$F_{2y} = +7824 \qquad F_{2x} = +18,300$$

Computer $= 7707$ \qquad Computer $= 18,500$

A_y is actual, not as in the free body diagram; therefore, a minus sign is placed before $m_{A_{2y}}$.

Appendix C

Segment 1 (Absolute)

$$F_{1y} = -W_1 - mA_{1y} + F_{2y}$$
$$= -2800 - 1515 + 7707$$
$$F_{1y} = +3392$$
Computer $= 3442$

$$F_{1x} = -mA_{1x} + F_{2x}$$
$$= -(-2200) + 18{,}500$$
$$F_{1x} = +20{,}700$$
Computer $= 20{,}721$

Segment 2 (Absolute)

$$M_2 = W_2 r_2 \cos\theta_2 - r_2(mA_{2y}\cos\theta_2 - mA_{2x}\sin\theta_2) - (k^2 - r^2)_2 m_2\alpha_2 = 0$$
$$M_2 - [(2276)(22.4)(.9)] - 22.4[(-10{,}100).9 - (-18{,}300).42] - [(26.6)^2 - (22.4)^2]2.32(-100) = 0$$
$$M_2 - 45{,}700 + 31{,}200 + 53{,}000 = 0$$
$$M_2 + 38{,}500 = 0$$
$$M_2 = -38{,}500 \qquad \text{Computer} = -38{,}931$$

Segment 1 (Absolute)

$$M_1 - W_1 r_1 \cos\theta_1 - r_1(mA_{1y}\cos\theta_1 - mA_{1x}\sin\theta_1) - (k^2 - r^2)_1 m_1\alpha_1 + F_{2y}l_1\cos\theta_1 - F_{2x}l_1\sin\theta_1 - M_2 = 0$$
$$M_1 - 2800(12.5).79 - 12.5[1515(.79) - (-2200)(-.615)] - [(15.5)^2 - (12.5)^2]2.858(-4.35) + 7707(28.6).79 - 18{,}500(28.6)(-.615) - (-38{,}500) = 0$$
$$M_1 - 27{,}600 + 1940 + 1042 + 174{,}000 + 325{,}000 + 38{,}500 = 0$$
$$M_1 = -512{,}882 \qquad \text{Computer} = -507{,}190$$

Segment 2 (Relative)

$$F_{2y} = -WT_2 + mr\omega^2\sin\theta_2 + m2V_2\omega_1\sin\theta_2 + mr\alpha_2\cos\theta_2 - mR\omega_1^2\sin\phi + mR\alpha_1\cos\phi$$
$$= -2276 + 2.32(22.4)(10.12)^2.42 + 2.32(2)(227)(8.63).42 + 2.32(22.4)(105.98).9 - 2.32(43.4)(8.63)^2.19 + 2.32(43.4)(4.35).98$$
$$= -2276 + 2240 + 3820 + 4950 - 1420 + 430$$
$$F_{2y} = +7744 \qquad \text{Computer} = 7707$$

$$F_{2x} = +mr\omega^2(.9)\cos\theta_2 + m2V_2\omega_1(.9)\cos\theta_2 - mr\alpha(.42)\sin\theta_2 + mR_1^2(.98)\cos\phi + mR\alpha_1(.19)\sin\phi$$
$$= +4800 + 8200 - 2320 + 7350 + 83$$
$$F_{2x} = 18{,}113 \qquad \text{Computer} = 18{,}500$$

Appendix C

Segment 1 (Relative)

$F_{1y} = WT_1 - mr\omega_1^2 \sin 38° + mr\alpha_1 \cos 38° + F_{2y}$
$\phantom{F_{1y}} = -2800 - 2.858(12.5)(8.63)^2(.615) + 2.858(12.5)(4.35).79 + 7707$
$\phantom{F_{1y}} = -2800 - 1635 + 123 + 7707$

$F_{1y} = +3395 \quad \text{Computer} = 3442$

$F_{1x} = mr\omega^2(.79) \cos 38° + mr\alpha(.615) \sin 38° + F_{2x}$
$\phantom{F_{1x}} = +2100 + 95 + 18,500$

$F_{1x} = -20,695 \quad \text{Computer} = 20\,721$

Segment 2 (Relative)

$M_2 - WT \cos \theta_2 r_2 + mk^2\alpha_2 - mR\omega_1^2 \sin 36° r_2 + mR\alpha_1 \cos 36° r_2 = 0$

$M_2 - 2276(.9)22.4 + 2.32(26.6)^2(105.98) - 2.32(43.4)(8.63)^2(.59)22.4$
$ + 2.32(43.4)(4.35)(.81)22.4 = 0$

$M_2 - 45,800 + 174,000 - 99,000 - 7950 = 0$

$M_2 + 37,150 = 0$

$M_2 = -37,150 \quad \text{Computer} = -38,931$

Segment 1 (Relative)

$M_1 - WT \cos \theta_1 r_1 + mk^2\alpha_1 + F_{2y} \cos 38° l_1 + F_{2x} \sin 38° l_1 - M_2 = 0$

$M_1 - 2800(.79)12.5 + 2.858(15.5)^2(4.35) + 7707(.79)28.6$
$ + 18,500(.615)28.6 - (-37,150) = 0$

$M_1 - 27,800 + 3000 + 174,000 + 325,000 + 37,150 = 0$

$M_1 = -511,350 \quad \text{Computer} = -507,190$

appendix D

BALL BOUNCE ON SOLID SURFACE FORMULAS

These formulas and their use are discussed in the section on ball spin in Chapter 7 (see pp. 84–88).

TEST FOR TWO-DIMENSIONAL NO SLIP CONDITION

$$\mu \geq \frac{V_f \cos \theta_f - V_i \cos \theta_i}{V_f \sin \theta_f + V_i \sin \theta_i}$$

Given: $e, r, \mu_s, \mu_d, V_i, \theta_i, \omega_i$; find: $V_f, \theta_f,$ and ω_f. ($\theta = 90°$ when $2\omega_i r = 5 V_i \cos \theta_i$.)

Solid Sphere

$(I = \tfrac{2}{5}mr^2)$

$$\omega_f = \frac{2\omega_i r - 5 V_i \cos \theta_i}{7r}$$

$$V_f = \tfrac{1}{7}\sqrt{(5 V_i \cos \theta_i - 2\omega_i r)^2 + (7 e V_i \sin \theta_i)^2}$$

$$\tan \theta_f = \frac{7 e V_i \sin \theta_i}{5 V_i \cos \theta_i - 2\omega_i r}$$

Thin Shelled Sphere

$(I = \tfrac{2}{3}mr^2)$

$$\omega_f = \frac{2\omega_i r - 3 V_i \cos \theta_i}{5r}$$

$$V_f = \tfrac{1}{5}\sqrt{(3 V_i \cos \theta_i - 2\omega_i r)^2 + (5 e V_i \sin \theta_i)^2}$$

$$\tan \theta_f = \frac{5 e V_i \sin \theta_i}{3 V_i \cos \theta_i - 2\omega_i r}$$

TEST FOR TWO-DIMENSIONAL SLIP CONDITION

The following formulas are used when μ static is exceeded ($I = \tfrac{2}{5} mr^2$ or $I = \tfrac{2}{3} mr^2$).

$$V_f = V_i \sqrt{e^2 \sin^2 \theta_i + [\cos \theta_i - \mu_d(1 + e) \sin \theta_i]^2}$$

$$\tan \theta_f = \frac{e \sin \theta_i}{\cos \theta_i - \mu_d(1 + e) \sin \theta_i}$$

When $I = \tfrac{2}{3} mr^2$:

$$\omega_f = \frac{2r\omega_i - 3\mu_d(1 + e) \sin \theta_i V_i}{2r}$$

When $I = \tfrac{2}{5} mr^2$:

$$\omega_f = \frac{2r\omega_i - 5\mu_d(1 + e) \sin \theta_i V_i}{2r}$$

TWO DIMENSION

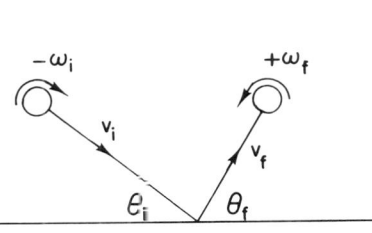

THREE DIMENSION

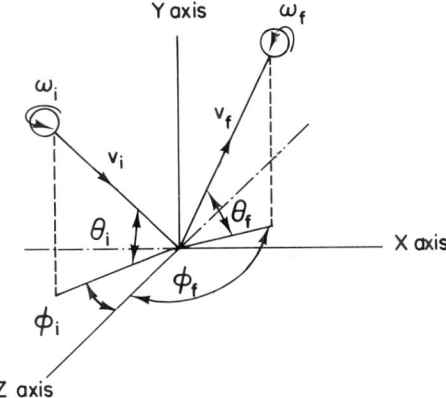

Fig. D-1
Two- and three-dimension diagrams for ball bounce formulas.

THREE-DIMENSIONAL NO SLIP CONDITION

When $I = \tfrac{2}{5} mr^2$:

$$\omega_{z_f} = \frac{2r\omega_{z_i} - 3V_i \cos \theta_i \sin \phi_i}{5r}$$

$$\omega_{x_f} = \frac{2r\omega_{x_i} - 3V_i \cos \theta_i \cos \phi_i}{5r}$$

$$\omega_{y_f} = \omega_{y_i}$$

$$(5V_f)^2 = 25e^2 V_i^2 \sin^2 \theta_i + (2r\omega_{x_i} - 3V_i \cos \theta_i \cos \phi_i)^2 + (2r\omega_{z_i} - 3V_i \cos \theta_i \sin \phi_i)^2$$

$$\tan \phi_f = \frac{2r\omega_{z_i} - 3V_i \cos \theta_i \sin \phi_i}{2r\omega_{x_i} - 3V_i \cos \theta_i \cos \phi_i}$$

$$\sin \theta_f = \frac{5eV_i \sin \theta_i}{\sqrt{25 e^2 V_i^2 \sin^2 \theta_i + (2r\omega_{z_i} - 3V_i \cos \theta_i \sin \phi_i)^2 + (2r\omega_{x_i} - 3V_i \cos \theta_i \cos \phi_i)^2}}$$

When $I = \frac{2}{5}mr^2$:

$$\omega_{z_f} = \frac{2r\omega_{z_i} - 5V_i \cos\theta_i \sin\phi_i}{7r}$$

$$\omega_{x_f} = \frac{2r\omega_{x_i} - 5V_i \cos\theta_i \cos\phi_i}{7r}$$

$$\omega_{y_f} = \omega_{y_i}$$

$$(7V_f)^2 = 49e^2V_i^2 \sin^2\theta_i + (2r\omega_{x_i} - 5V_i \cos\theta_i \cos\phi_i)^2 + (2r\omega_{z_i} - 5V_i \cos\theta_i \sin\phi_i)^2$$

$$\tan\phi_f = \frac{2r\omega_{z_i} - 5V_i \cos\theta_i \sin\phi_i}{2r\omega_{x_i} - 5V_i \cos\theta_i \cos\phi_i}$$

$$\sin\theta_f = \frac{7eV_i \sin\theta_i}{\sqrt{49e^2V_i^2 \sin^2\theta_i + (2r\omega_{z_i} - 5V_i \cos\theta_i \sin\phi_i)^2 + (2r\omega_{x_i} - 5V_i \cos\theta_i \cos\phi_i)^2}}$$

THREE-DIMENSIONAL SLIP CONDITION

When $I = \frac{2}{3}mr^2$:

$$\omega_{z_f} = \frac{2r\omega_{z_i} - 3\mu_D V_i \sin\theta_i \cos\phi_i(1 + e)}{2r}$$

$$\omega_{x_f} = \frac{2r\omega_{x_i} - 3\mu_D V_i \sin\theta_i \cos\phi_i(1 + e)}{2r}$$

$$\omega_{y_f} = \omega_{y_i}$$

$$V_f = V_i\sqrt{e^2 \sin^2\theta_i + (\cos\theta_i - \mu_D(1 + e)\sin\theta_i)^2}$$

$$\tan\phi_f = \frac{\cos\theta_i \sin\phi_i - \mu_D \sin\theta_i \sin\phi_i(1 + e)}{-\cos\theta_i \cos\phi_i + \mu_D \sin\theta_i \cos\phi_i(1 + e)}$$

$$\sin\theta_f = \frac{e \sin\theta_i}{\sqrt{e^2 \sin^2\theta_i + [\cos\theta_i - \mu_D \sin\theta_i(1 + e)]^2}}$$

When $I = \frac{2}{5}mr^2$:

$$\omega_{z_f} = \frac{2r\omega_{z_i} - 5\mu_D V_i \sin\theta_i \cos\phi_i(1 + e)}{2r}$$

$$\omega_{x_f} = \frac{2r\omega_{x_i} - 5\mu_D V_i \sin\theta_i \cos\phi_i(1 + e)}{2r}$$

$$\omega_{y_f} = \omega_{y_i}$$

V_f, $\tan\phi$, and $\sin\theta$ are the same as for $I = \frac{2}{3}mr^2$.

appendix E
RECOMMENDED COURSE OUTLINE

At the University of Massachusetts, the course outlines presented here are followed, in the main, in one undergraduate and three graduate courses. This text supports the teaching of all four courses.

UNDERGRADUATE LABORATORY COURSE

1. Review of math and physics needed for problems related to human motion — Appendix A, Problems 1–11. (Two to three weeks.)

2. A. Determine the resultant of three vectors using three spring scales. Work similar to Problem 5, Appendix A.
   B. Determine the total body center of gravity of a selected body position using a board with a weighing scale under one end. See Williams and Lissner (H 29, page 137).
   C. Determine the total body center of gravity using measurements only. Problem 13, Appendix A. (One week.)

3. Determine the force on the wire, the X and Y forces of the wood against the wall, the resultant direction of the force of the end of the wood against the wall using two methods, and the magnitude of the resultant force. (Fully covered in Chapter 4, illustrated in Fig. 4–3.) (One week.)

4. Review the planes of the joint centers and determine each body segment weight using the water displacement method (see Chapter 3). Only a demonstration is possible during the lab time since only one or two water tanks are generally available. (One week.)

5. A. Demonstrate the technique of tracing motion pictures frame by frame to obtain a composite of the whole motion. Show the motion relative to the earth and relative to a moving fixed point (swimming, rowing).
   B. Demonstrate the use of stroboscopic equipment to obtain a composite picture of a whole motion. (One week.)

6. Draw the displacement, velocity, and acceleration curves for a one-segment motion. Each person determines the velocity for a different body position after the displacement curve is drawn, and the results of the team effort are used to draw the velocity curve. Each person then determines the acceleration for a given position and the total data are used to draw the acceleration curve. (One week.)

7. Demonstrate the use of electromyography and electrogoniometry. (One or two weeks.)

8. Determine the coefficient of friction of two objects using two methods (incline board and table top). Work the sliding friction problem, Problem 12, Appendix A. (One week.)

9. A. Determine the coefficient of restitution of several balls at various drop heights.
   B. Using a spring loaded trajectory apparatus (found in most physics labs), determine the range of a shot mathematically and compare with the measured distance (see Problem 8, Appendix A). (One week.)

10. A. Determine the moment of inertia of a pendulum. Determine the radius of gyration and center of percussion of a baseball bat, tennis racket, and golf club.
    B. Do centripetal force problem with apparatus found in physics labs (see Problem 14, Appendix A). (One week.)

GRADUATE STUDENT COURSES

First Course

1. Review the force and moment formulas and continue drawing the free body diagrams for two- and three-segment motions.

2. Take a still picture of a symmetrical, planar position and perform a whole body force and moment analysis. Obtain the segment weights and lengths of the subject analyzed.

3. Combine the stroboscope and elgon techniques to obtain joint angular changes of a motion and compare the results.

4. Take motion pictures of a two-segment motion (upper arm, and forearm plus hand), with the EMG recording the biceps, triceps, and brachioradialis. Use the computer program (Appendix B), and compare the force data obtained with the EMG data.

5. Trace a symmetrical, whole body motion, key punch the necessary data, and use the computer program (Appendix B). Interpret the output data.

Second Course

1. Take motion pictures of a three-segment pendulum motion, perform the frame by frame tracing, and input the necessary data using the computer program (Appendix B). Then draw the displacement, velocity, and acceleration curves by hand (graph paper and french curve), and compare the answers with those obtained from the computer program. Also study the changes of the curve fit of the computer program by using several polynomial numbers. Choose one instantaneous position of the pendulum and calculate the forces and moments for all three segments by hand. (Draw free body diagrams, insert proper data into formulas, and determine answers using a slide rule.) Compare with the computer output.

2. Photograph a motion and determine only the path of the total body center of gravity. Choose a nonsymmetrical motion so the tracing will require reducing the body to a single-chain link system. This is done by connecting the points mid-way between the joint centers (ankle-ankle, knee-knee, etc.), resulting in a one leg–one arm man with weights equivalent to the two legs and arms. The computer program presented can then determine the total body center of gravity. Only the pertinent portion dealing with center of gravity can be extrapolated for those wishing to do considerable work in this area.

3. Photograph, trace, and obtain the forces and moments (using the computer program, Appendix B) of a nonsymmetrical motion that does not involve any external forces. Interpret the results.

Third Course

1. Analyze the changing position of the trunk center of gravity due to extreme flexion and extension, shoulder movements, and nonplanar positions. This must be done by hand using the cardboard-lead method. See Chapter 3.

2. Photograph and analyze a motion that has no fixed point (any in water or in air motion).

3. Analyze a motion that involves a continuously changing external force (rowing, canoeing, kayaking, wall pulley exercises, swim cable exercises, etc.).

4. Analyze a motion that involves a force of impact. Determine the force of impact and apply this force at the appropriate position.

5. Write the necessary computer programs for the ball spin formulas presented in Chapter 7. Obtain the coefficient of friction and restitution for any chosen sport conditions and use the programs.

BIBLIOGRAPHY

A: METHODS OF RECORDING DATA

1. BASMAJIAN, J. V. and A. LATIF. 1957. Integrated Actions and Functions of the Chief Flexors of the Elbow: a detailed electromyographic analysis. *J. Bone Jt. Surg.* 39: 1106–18.
2. BASMAJIAN, J. V. 1959. Spurt and Shunt Muscles: an electromyographic confirmation. *J. Anat.* 93: 551–53.
3. ———. 1962. *Muscles Alive — Their Functions Revealed by Electromyography.* Baltimore. Williams & Wilkins.
4. ——— and ANTHONY TRAVILL. 1961. Electromyography of the Pronator Muscles in the Forearm. *Anat. Rec.* 139: 45–49.
5. BASMAJIAN, J. V. 1957. Electromyography of Two-Joint Muscles. *Anat. Rec.* 129: 371–80.
6. BEIRMAN, W. A. 1957. A New Apparatus: a method for the measurement of minimal muscle force. *Arch. Phys. Med.* 38: 450.
7. BIERMAN, H. R. and V. R. LARSEN. 1946. Reaction of the Human to Impact Forces Revealed by High Speed Motion Picture Technique. *J. Aviat. Med.* 17: 407–12.
8. BIERMAN, W. and L. J. YAMSHON. 1948. Electromyography in Kinesiologic Evaluations. *Arch. Phys. Med.* 29: 206–11.
9. BROUHA, LUCIEN. 1960. *Physiology in Industry.* New York: Pergamon Press.
10. CHESTERMAN, W. D. 1951. *The Photographic Study of Rapid Events.* London: Oxford University Press.
11. COOPER, JOHN M. 1958. Mechanics of Human Movement — Throwing. *61st Ann. Proc. Coll. Phys. Educ. Assoc.*
12. COURTNEY-PRATT, J. S. 1949. A New Method for the Photographic Study of Fast Transient Phenomena. *Research* 2: 237.

13. CURETON, T. K. 1939. Elementary Principles and Techniques of Cinematographic Analysis. *Res. Quart.* 10: 3–24.
14. DAVIS, J. F. 1959. Manual of Surface Electromyography. *WADC Tech. Report:* 59–184.
15. DAVIS, P. R., J. D. G. TROUP and J. H. BURNARD. 1965. Movements of the Thoracic and Lumbar Spine when Lifting: a chronocyclophotographic study. *J. Anat.* 99: 13–26.
16. DRILLIS, R. J. 1959. The Use of Gliding Cyclograms in Biomechanical Analysis of Movement. *Human Factors* 1: 2.
17. EDGERTON, H. E. and R. S. CARLSON. 1950. The Stroboscope as a Light Source for Motion Pictures. *SMPTE Journal* 55: 88–100.
18. EDGERTON, H. E., J. K. GERMESHAUSEN, and H. E. GRIER. 1937. High Speed Photographic Methods of Measurement. *J. Appl. Phys.* 8: 2.
19. EDGERTON, H. E and J. R. KILLIAN. 1954. *Flash Ultra-High Speed Photography.* Boston: Branford.
20. EDGERTON, H. E. and C. W. WYCKOFF. 1951. A Rapid Action Shutter with No Moving Parts. *J. Soc. Mot. Tele. Eng.* 56: 398.
21. FINLEY, F. R. and P. V. KARPOVICH. 1964. Electrogoniometric Analysis of Normal and Pathological Gaits. *Res. Quart.* 35: 379–84.
22. FINLEY, F. R. and R. W. WIRTA. Myocoder-Computer Study of Electromyographic Patterns. *Arch. Phys. Med. Rehabil.* 48: Jan. 1967.
23. FLOYD, W. F. and P. SILVER. 1950. Electromyographic Study of Patterns of Activity of the Anterior Abdominal Wall in Man. *J. Anat.* 84: 132–45.
24. FOEPPL, L. 1949. Slow-Motion Pictures of Impact Tests by Means of Photo-elasticity. *J. Appl. Mech.* 16: 173.
25. FREEDMAN, L. and R. R. MUNRO. 1966. Abduction of the Arm in the Scapular Plane: scapular and glenohumeral movements. A roentgenographic study. *J. Bone Jt. Surg.* 48: 1503–10.
26. FUSFELD, H. I. and J. C. FEDER. 1950. Study of Deformation at High Strain Rates Using High Speed Motion Pictures. *Amer. Soc. Test. Mater.* 42.
27. GARRETT, R. E., C. J. WIDULE, and G. E. GARRETT. 1968. Computer Aided Analysis of Human Motion. *Kines. Rev. AAHPER*, Washington, D. C.
28. GOLDMAN, D. E. and J. B. BOORSTIN. 1964. X-Ray Cinematographic Observations of the Motions of Animals Exposed to Mechanical Vibration. *Aero. Med.*, 35th Ann. Sci. Meeting, May.
29. GOLLNICK, PHILIP D. and PETER V. KARPOVICH. 1964. Electrogoniometric Study of Locomotion and of Some Athletic Movements. *Res. Quart.* 35: 357–69.
30. *Handbook of High Speed Photography.* 1963. West Concord, Mass.: General Radio Co.
31. HOUTZ, S. J. and F. J. FISCHER. 1959. An Analysis of Muscle Action and Joint Excursion During Exercise on a Stationary Bicycle. *J. Bone Jt. Surg.* 41-A: 123–31.
32. HYZER, WILLIAM G. 1963. *Engineering and Scientific High Speed Photography.* New York: Macmillan.
33. INMAN, V. T., H. J. RALSTON, J. B. SAUNDERS, B. FEINSTEIN, and E. W. WRIGHT. 1952. Relation of Human Electromyogram to Muscular Tension. *EEG Clin. Neurophysiol.* 4: 187.
34. JONES, G. A. 1952. *High Speed Photography.* New York: John Wiley.
35. JOSEPH, J. 1960. *Man's Posture, Electromyographic Studies.* Springfield, Ill.: Charles C Thomas.
36. ——— and R. WATSON. 1967. Telemetering Electromyography of Muscles Used in Walking up and down Stairs. *J. Bone Jt. Surg.* 49-B: 774–80.

37. Karpovitch, P. V., R. A. Weiss, and E. R. Elbel. 1946. Relation Between Leg-Lift and Sit-Up. *Res. Quart.* 17: 21.
38. Karpovitch, P. V. 1950. A Frictional Bicycle Ergometer. *Res. Quart.* 21: 210.
39. ——— and G. P. Karpovitch. 1959. Electrogoniometer: a new device for study of joints in action. *Fed. Proc.* 18: 79.
40. Karpovitch, P. V. and L. B. Wilklow. 1959. A Goniometric Study of the Human Foot in Standing and Walking. *U. S. Armed Froces Med. J.* 10: 885–903.
41. ———. 1960. Goniometric Study of the Human Foot in Standing and Walking. *Indus. Med. Surg.* 29: 338.
42. Karpovitch, P. V., E. L. Herden, and M. M. Asa. 1960. Electrogoniometric Study of Joints. *U. S. Armed Forces Med J.* 11: 424.
43. Keagy, R. D., J. Brumlik and J. J. Bergan. 1966. Direct Electromyography of Psoas Major in Man. *J. Bone Jt. Surg.* 48A: 1377–82.
44. Levens, Alexander S. 1965. *Graphical Methods in Research.* New York: John Wiley.
45. Licht, Sidney, ed. 1961. *Electrodiagnosis and Electromyography*, 2nd ed. New Haven: Elizabeth Licht, Publ.
46. Lockhart, Aileene. 1941. A Survey of Devices Used in Measuring Short Time Intervals. *Res. Quart.* 12: 757–64.
47. Mascelli, Joseph V. and Arthur Miller. 1966. *American Cinematographer Manual.* Hollywood, Cal.: Amer. Society of Cinematographers Holding Corp.
48. Nelson, Richard C., Kenneth L. Petak, and Gary S. Pechar. 1969. Use of Stroboscopic-Photographic Techniques in Biomechanics Research. *Res. Quart.* 40: 424–26.
49. Norris, F. H. 1963. *The EMG, A Guide and Atlas for Practical Electromyography.* New York: Grune and Stratton.
50. O'Connell, A. L. and O. A. Mortensen. 1957. An Electromyographic Study of the Leg Musculature During Movements of the Free Foot and During Standing. *Anat. Rec.* 127: 342.
51. Plagenhoef, S. C. 1968. Gathering Kinesiological Data Using Modern Measuring Devices. *J. A. A. H. P. E. R.*, Oct.: 81–82.
52. Rebikoff, Dimitri and Paul Cherney. 1955. *A Guide to Underwater Photography.* New York: Greenberg.
53. Reid, Stephen E., Joseph A. Tarkington, and Thomas E. Healion. 1963. Medical Telemetry in Sports. *Proc. Fifth Nat. Conf. Med. Asp. Sports*, Portland, Oregon.
54. Ricci, Benjamin. 1967. *Physiological Basis of Human Performance.* Philadelphia: Lea & Febiger.
55. Roberts, V. L. 1966. Strain-Gauge Techniques in Biomechanics. *Exp. Mech.*, March: 1–4.
56. Scheving, Lawrence E. and John E. Pauly. 1959. An Electromyographic Study of Some Muscles Acting on the Upper Extremity of Man. *Anat. Rec.* 135: 239–45.
57. Schwartz, R. P., A. L. Heath, and J. W. Wright. 1934. Kinetics of the Human Gait. *J. Bone Jt. Surg.* 16: 343.
58. Sigerseth, P. O. and C. H. McCloy. 1955. Electromyographic Study of Selected Muscles Involved in Movements of Upper Arm at Scapulohumeral Joint. *Res. Quart.* 27: 409–17.
59. Slaughter, Duane R. 1959. Electromyographic Studies of Arm Movements. *Res. Quart.* 30: 326–37.
60. Society of Motion Picture and Television Engineers. *High Speed Photography* 1–6: 1952–57. New York.

61. SULLIVAN, W. E., O. A. MORTENSEN, M. MILES, and L. S. GREENE. 1950. Studies of M. Biceps Brachii During Normal Voluntary Movement at the Elbow. *Anat. Rec.* 107: 243–51.
62. TRAVILL, ANTHONY and JOHN V. BASMAJIAN. 1960. Electromyography of the Supinators of the Forearm. *Anat. Rec.* 139: 557–60.
63. TUTTLE, W. W. and A. J. WENDLER. 1945. The Construction, Calibration and Use of an Alternating Current Electrodynamic Brake Bicycle Ergometer. *J. Lab. Clin. Med.* 30: 173–83.
64. VREELAND, R. W., D. H. SUTHERLAND, J. J. DORSA, L. A. WILLIAMS, C. C. COLLINS, and E. R. SCHOTTSTEADT. 1961. A Three-Channel Electromyograph with Synchronized Slow-Motion Photography. IRE Trans. *Bio-Med. Electr.* 8: 4–6.
65. WADDELL, J. H. and J. W. WADDELL. 1955. *Photographic Motion Analysis.* Chicago: Indust. Lab. Publ.
66. WATERLAND, JOAN C. and GEORGIA M. SHAMBES. 1969. Electromyography: one link in the experimental chain of kinesiological research. *J. Amer. Phys. Ther. Assn.* 49: 1351–56.
67. WILSON, G. D. and W. H. STASCH. 1945. Photographic Record of Joint Motion. *Arch. Phys. Med.* 26: 361.
68. YAMSHON, L. F. and W. BIERMAN. 1949. Kinesiologic Electromyography. III. The Deltoid. *Arch. Phys. Med.* 30: 286–89.
69. ———. 1948. Kinesiologic Electromyography. II: The Trapezius. *Arch. Phys. Med.* 29: 647–51.
70. Zankel, H. T. 1951. Photogoniometry. *Arch. Phys. Med.* 32: 227.

B: CENTER OF GRAVITY

1. BARTER, JAMES T. 1957. Estimation of the Mass of Body Segments. *WADC Tech. Report:* 57–260.
2. BEHNKE, A. R., B. G. FEEN, and W. C. WELHAM. 1942. The Specific Gravity of Healthy Men. *J. A. M. A.* 118: 495–98.
3. BOYD, E. 1933. The Specific Gravity of the Human Body. *Human Biol.* 5: 646–72.
4. BRAUNE, W. and O. FISCHER. 1889. Über den Schwerpunkt des menschlichen Körpers, mit Rücksicht auf die Ausrüstung des deutschen Infanteristen. *Abh. d. math. phys. cl. d. k. Sächs. Gesellsch. d. Wiss.* 15: 561–72.
5. CLAUSER, CHARLES E., JOHN T. MCCONVILLE, and J. W. YOUNG. 1969. Weight, Volume, and Center of Mass of Segments of the Human Body. Wright Air Development Center, AMRL-TR-69-70, Wright-Patterson Air Force Base, Ohio.
6. CLEAVELAND, HENRY G. 1955. The Determination of the Center of Gravity in Segments of the Human Body. Master's Thesis, University of California.
7. DEMPSTER, W. T. 1955. Space Requirements of the Seated Operator. *WADC Tech. Report:* 55–159.
8. DuBOIS, J. and W. R. SANTSCHI. 1963. *The Determination of the Moment of Inertia of Living Human Organism.* New York: John Wiley.
9. DUGGAR, S. C. 1963. The Center of Gravity of the Human Body. *J. Hum. Fact. Soc.* 4: 131–48.
10. DUPERTUIS, C. W., G. C. PITTS, E. F. OSSERMAN, W. C. WELHAM, and A. R. BEHNKE. 1951. Relation of Specific Gravity to Body Build in a Group of Healthy Men. *J. Appl. Physiol.* 3: 676–80.
11. FISCHER, O. 1906. *Theoretische Grundlagen fur eine Mechanik der lebenden Korper.* Berlin: B. G. Teubner.
12. HIRT, SUSANNE E., CORRINE FRIES, and FRANCES A. HELLEBRANDT. 1944. Center of Gravity of the Human Body. *Arch. Phys. Therapy* 25: 280.

13. KJELDSEN, K. 1969. Body Segment Weights of College Women. Unpublished Master's Thesis, University of Massachusetts.
14. KROGMAN, W. M. and F. E. JOHNSTON. 1963. Human Mechanics: Four Monographs Abridged. *AMRL Tech. Doc. Report* 63–123.
15. PALMER, CARROLL E. 1928. Center of Gravity of the Human Body During Growth. 1. An improved apparatus for determining the center of gravity. *Amer. J. Phys. Anthro.* 11: 423–55.
16. ———. 1944. Studies of the Center of Gravity in the Human Body. *Child Devel.* 15: 99.
17. PARKS, JESSE L., JR. 1959. An Electromyographic and Mechanical Analysis of Selected Abdominal Exercises. Unpublished Doctoral Thesis, University of Michigan
18. PLAGENHOEF, STANLEY. 1966. Methods for Obtaining Kinetic Data to Analyze Human Motions. *Res. Quart.* 37: 103–12.
19. SANTSCHI, W. R., J. DUBOIS, and C. OMOTO 1963. Moments of Inertia and Centers of Gravity of the Living Human Body. *AMRL Tech. Doc. Report* 63–36.
20. SWEARINGEN, J. J. 1953. Determination of Centers of Gravity. Dept. Comm. Civ. Aer. Med. Res. Lab. Proj. 53–203.
21. ———. 1962. Determination of Centers of Gravity of Man. *F. A. A. Report* 62–14.
22. WEINBACK, A. P. 1938. Contour Maps, Center of Gravity, Moment of Inertia and Surface Area of the Human Body. *Human Biol.* 10: 356.
23. WELHAM, W. C. and A. R. BEHNKE. 1942. The Specific Gravity of Healthy Men. *J. A. M. A.* 118: 498–501.

C: PHYSICS AND MECHANICS

1. "A Boom with a Bounce." December 3, 1965. *Life* Magazine.
2. *American Institute of Physics Handbook.* New York: McGraw-Hill.
3. BARNES, GEORGE. 1958. Study of Collisions. *Amer. J. Phys.* 26: 5.
4. BAYES, JANE H. and WILLIAM T. SCOTT. 1963. Billiard Ball Collision Experiment. *Amer. J. Phys.* 31: 197.
5. BIKERMAN, JACOB J. 1941. Friction and Adhesion. *Phil. Mag.* 7: 67–76.
6. BOWDEN, F. P. and D. TABOR. 1950. The Friction and Lubrication of Solids. Oxford: Clarendon Press.
7. BOWDEN, F. P. and E. H. FREITAG. 1958. The Friction of Solids at Very High Speeds. *Proc. Roy. Soc.* 248: 350–67.
8. BURGEL, B. 1967. Centrifugal Force. *Amer J. Phys.* 35: 649–50.
9. BURWELL, J. T. and E. RABINOWICZ. 1953. The Nature of the Coefficient of Friction. *J. Appl. Phys.* 24: 136–39.
10. GROVES, GORDON W. 1967. Acceleration Referred to Moving Curvilinear Coordinates. *Amer. J. Phys.* 10: 927–29.
11. HALLIDAY, D. and R. RESNICK. 1965. *Physics for Students of Science and Engineering*, Part I. New York: John Wiley.
12. HAM, C. W. and E. J. CRANE. 1948. *Mechanics of Machinery.* New York: McGraw-Hill.
13. HAWKINS, R. R., ed. 1958. Scientific Medical and Technical Books, 2nd ed. Washington: Nat. Res. Council.
14. HEGARTY, JOHN C. 1966. Student Experiment on Coriolis Force. *Amer. J. Phys.* 34, No. 2 (Feb 1966), 157–59.
15. HEMMING, GEORGE W. 1899. *Billiards Mathematically Treated.* London: Macmillan.

16. KUO, SHAN S. 1965. *Numerical Methods and Computers*. Pp. 234–35. Reading, Mass.: Addison-Wesley.
17. LONEY, S. L. 1950. *Dynamics of a Particle and of Rigid Bodies*. Cambridge: Cambridge University Press.
18. ———. 1951. *Solutions of the Examples of Dynamics of a Particle and of Rigid Bodies*. Cambridge: Cambridge University Press.
19. MANEY, CHARLES A. 1952. Experimental Study of Sliding Friction. *Amer. J. Phys.* 20: 203.
20. MARTIN, R. C. and W. JETT. 1963. *Guide to Scientific and Technical Periodicals*. Chicago: Swallow Press.
21. MCDONALD, JAMES E. 1952. The Coriolis Effect. *Sci. Amer.*, May: 72.
22. MCLEAN, W. G. and E. W. NELSON. 1952. *Engineering Mechanics*. New York: Schaum Publ.
23. MELLEN, WALTER R. 1968. Superball Rebound Projectiles. *Amer. J. Phys.* 36: 845.
24. MERIAM, J. L. 1952. *Mechanics*. Part I — Statics. New York: John Wiley.
25. ———. 1966. *Mechanics*. Part II — Dynamics. New York: John Wiley.
26. MINER, DOUGLAS F. and JOHN B. SEASTONE, eds. 1955. *Handbook of Engineering Materials*. New York: John Wiley.
27. MORIN, ARTHUR. 1834. Nouvelles Experiences Sur le Frottement. Inst. De France Acad. Roy. Des Sci.
28. OSGOOD, WILLIAM F. 1937. *Mechanics*. New York: Macmillan.
29. PALMER, FREDERIC. 1949. What About Friction? *Amer. J. Phys.* 17: 181.
30. ———. 1951. Friction. Sci. Amer., February.
31. PEARLMAN, N. 1967. Vector Representation of Rigid-Body Rotation. *Amer. J. Phys.* 35: 1164.
32. Physical Science Study Committee. 1965. *Physics*. Boston: D. C. Heath.
33. POYNTING and THOMSON. 1922. *Textbook of Physics*. Volume 1. London: Griffin.
34. PRESCOTT. 1923. *Mechanics of Particles and of Rigid Bodies*. 2nd ed. London: Longmans, Green.
35. RABINOWICZ, ERNEST. 1956. Stick and Slip. *Sci. Amer.*, May.
36. ———. 1965. *Friction and Wear of Materials*. New York: John Wiley.
37. ROUSE, HUNTER. 1946. *Elementary Mechanics of Fluids*. New York: John Wiley.
38. ROUTH, E. J. 1877. *Dynamics of a System of Rigid Bodies*. London: Macmillan.
39. SEARS, F. W. 1958. *Mechanics, Heat, and Sound*. Reading, Mass: Addison-Wesley.
40. SHAMES, I. H. 1966. *Engineering Mechanics. Dynamics*, Volume II. Englewood Cliffs, N. J.: Prentice-Hall.
41. STREETER, V. L. 1966. *Fluid Mechanics*, 4th ed. New York: McGraw-Hill.
42. STODDARD, J. T. 1913. *The Science of Billiards*. Boston: Butterfield.
43. SUTTON, RICHARD M. 1955. Two Notes on the Physics of Walking. *Amer. J. Phys.* 23: 490.
44. SYNGE, J. L. and B. A. GRIFFITH. 1949. *Principles of Mechanics*. New York: McGraw-Hill.
45. THURNAUR, PETER G. 1967. Kinematics of Finite, Rigid-Body Displacements. *Amer. J. Phys.* 35: 1145–54.
46. TIMOSHENKO, S. 1934. *Theory of Elasticity*. New York: McGraw-Hill.
47. ——— and D. H. YOUNG. 1956. *Engineering Mechanics*. New York: McGraw-Hill.
48. VAN NAME, F. W., JR. 1958. Experiment for Measuring the Coefficient of Restitution. *Amer. J. Phys.* 26: 386.
49. *Van Nostrand's Scientific Encyclopedia*. 1958. New York: Van Nostrand Reinhold.

D: FREE BODY DIAGRAMS AND COMPUTERIZED MECHANICAL ANALYSIS

1. Bartholomew, S. H. 1952. Determination of Knee Moments During the Swing Phase of Walking and Physical Constants of the Human Shank. Series II. Pros. Dev. Res. Proj. Inst. Eng. Res., Univ. of Calif., Berkeley.
2. Bresler, B. and J. P. Frankel. 1950. The Forces and Moments in the Leg During Level Walking. *Trans. Amer. Soc. Mech. Eng.* 72: 27.
3. Calkins, Gerald. 1970. A Computerized Extension of the Theory of Mechanical Analysis of Human Movement to Consider the Work Done at the Body Joints and its Relation to Energy Cost. Unpublished Doctoral Dissertation, University of Massachusetts.
4. Chaffin, Donald B. 1969. A Computerized Biomechanical Model — Development of and Use in Studying Gross Body Actions. *J. Biomechanics* 2: 429–41.
5. Dempster, Wilfrid T. 1958. Analysis of Two-Handed Pulls Using Free Body Diagrams. *J. Appl. Phys.* 13.
6. ———. 1955. Space Requirements of the Seated Operator: Geometrical, Kinematic, and Mechanical Aspects of the Body with Special Reference to the Limbs. *WADC Tech. Report:* 55–159.
7. ———. 1961. Free Body Diagrams as an Approach to the Mechanics of Human Posture and Motion. *Biomechanical Studies of the Musculo-Skeletal System*, F. G. Evans, ed. Springfield, Ill.: Charles C Thomas.
8. Plagenhoef, Stanley C. 1966. Methods for Obtaining Kinetic Data to Analyze Human Motions. *Res. Quart.* 37: 103–12.
9. ———. 1968. Computer Programs for Obtaining Kinetic Data on Human Movement. *J. Biomechanics* 1: 221–34.
10. Radcliffe, C. W. 1950. Information Useful in the Design of Damping Mechanisms for Artificial Knee Joints. Series III. Pros. Dev. Res. Proj. Inst. Eng. Res. Berkeley: Univ. of Calif.

E: ANTHROPOMETRY — SOMATOTYPE

1. Alexander, M. R. S. Zeigen, and I. Emanuel. 1961. Anthropometric Data Presented in Three-Dimensional Forms. *Amer. J. Phys. Anthro.* 19: 147–57.
2. Brozek, J. 1961. Techniques for Measuring Body Composition. Monograph, Washington, D. C.: Nat. Acad. Sci.
3. Churchill, E. and Katherine Bernhardi. 1957. WAF Trainee Body Dimensions: A Correlation Matrix. *WADC Tech. Rep.* 57–197.
4. Damon, Albert H. et al. 1966. The Human Body in Equipment Design. Cambridge, Mass.: Harvard University Press.
5. Damon, Albert H., H. Stoudt, and R. McFarland. 1963. *Human Engineering Guide to Equipment Design.* New York: McGraw-Hill. Chap. 11.
6. Daniels, G. S., H. C. Meyers, and E. Churchill. 1953. Anthropometry of Male Basic Trainees. *WADC Tech. Report:* 53.
7. ———, ———, and S. H. Worrall. 1953 Anthropometry of *WAF* Basic Trainees. *WADC Tech Report:* 53.
8. ———, E. Churchill, and H. Hertzberg. 1954. Anthropometry of Flying Personnel, 1950. *WADC Tech. Report:* 52.
9. Darcus, H. D. 1954. *The Range and Strength of Joint Movement.* Symposium on Human Factors in Equipment Design. London: H. K. Lewis.

10. DEMPSTER, WILFRID T., LAWRENCE A. SHERR, and JUDITH G. PRIEST. 1964. Conversion Scales for Estimating Humeral and Femoral Lengths and the Lengths of Functional Segments in the Limbs of American Caucasoid Males. *Human Biology* 36.

11. DUPERTUIS, C. W. and J. M. TANNER. 1950. The Pose of the Subject for Photogrammetric Anthropometry with Special Reference to Somatotyping. *Amer. J. Phys. Anthro.* 10: 331–51.

12. GAVAN, J. A. et al. 1951. Photography: an Anthropometric Tool. *Amer. J. Phys. Anthro.* 8: 27–48.

13. GENOVÉS, SANTIAGO. 1967. Proportionality of the Long Bones and their Relation to Stature among Meso-Americans. *Amer. J. Phys. Anthro.* 26: 67.

14. GEOGHEGAN, B. 1953. The Determination of Body Measurements, Surface Area and Body Volume by Photography. *Amer. J. Phys. Anthro.* 11: 97–120.

15. HEATH, BARBARA H. and J. E. L. CARTER. 1966. A Comparison of Somatotype Methods. *Amer. J. Phys. Anthro.* 24: 87.

16. ———. 1967. A Modified Somatotype Method. *Amer. J. Phys. Anthro.* 27:57.

17. HELLEBRANDT, FRANCES A. et al. 1938. The Location of the Cardinal Anatomical Orientation Planes Passing Through the Center of Weight in Young Adult Women. *Amer. J. Physiol.* 121: 465.

18. HERTZBERG, H. T. E. 1948. Post-War Anthropometry in the Air Force. *Amer. J. Phys. Anthro.* 6: 363–71.

19. ———, G. S. DANIELS, and E. CHURCHILL. 1954. Anthropometry of Flying Personnel, 1950. *WADC Tech. Report* 52: 321.

20. HERTZBERG, H. T. E. 1955. Some Contributions of Applied Anthropology to Human Engineering. *Ann. N. Y. Acad. Sci.* 63: 616–29.

21. ———. 1968. The Conference on Standardization of Anthropometric Techniques and Terminology. *Amer. J. Phys. Anthro.* 28: 1.

22. JONES, PETER R. and PATRICK G. STONE. 1964. An Advance in Somatotype Photography. *Amer. J. Phys. Anthro.* 22: 259.

23. JORDAN, JAMES H. 1969. Physiological and Anthropometrical Comparisons of Negroes and Whites. *J.A.H.P.E.R.*, Nov.-Dec.

24. KEENEY, C. E. 1955. Relationship of Body Weight to Strength Body Weight Ratio in Championship Weightlifters. *Res. Quart.* 26: 54–59.

25. KRAKOWER, H. 1935. Skeletal Characteristics of the High Jumper. *Res. Quart.* 6: 75–84.

26. ———. 1941. Skeletal Symmetry and High Jumping. *Res. Quart.* 12: 218–27.

27. KROLL, W. 1954. An Anthropometrical Study of Some Big Ten Varsity Wrestlers. *Res. Quart.* 25: 307–12.

28. LEWIS, A. S. 1969. Physique as a Determinant of Success in Sport with Particular Reference to Olympic Oarsmen. New Zealand *J.H.P.E.R.* 2: 5–21.

29. LITTIG, LAWRENCE W. 1964. Lens-Subject Distance for Photogrammetric Anthropometry. *Amer. J. Phys. Anthro.* 22: 1.

30. LOWRANCE, E. W. and H. B. LATIMER. 1957. Weights and Linear Measurements of 105 Human Skeletons from Asia. *Amer. J. Anat.* 101: 445–59.

31. MITCHEM, JOHN C. and BARBARA KAY ARSENAULT. 1960. An Evaluation of Anthropometric Studies Appearing in the Research Quarterly from 1940 to 1960. *Res. Quart.* 37: 438–40.

32. MONTAGU, M. F. A. 1960. *A Handbook of Anthropometry.* Springfield, Ill.: Charles C Thomas.

33. MORGAN, CLIFFORD T., ALPHONSE CHAPANIS, JESSE S. COOK III, and MAX W. LUND. 1963. *Human Engineering Guide to Equipment Design.* New York: McGraw-Hill.
34. ROEBUCK, J. A. 1952. Anthropometric Data: Human Body Dimensions and Weights. Douglas Aircraft Co., Inc.
35. SHELDON, W. H., C. W. DUPERTUIS and E. McDERMOTT. 1954. *Atlas of Men: A guide for somatotyping the adult male at all ages.* New York: Harper & Row.
36. SHELDON, W. H., S. S. STEVENS, and W. B. TUCKER. 1940. *The Varieties of Human Physique.* New York: Harper & Row.
37. SILLS, F. D. 1950. A Factor Analysis of Somatotyping and of their Relationship to Achievement in Motor Skills. *Res. Quart.* 21: 424–37.
38. ——— and P. W. EVERETT. 1953. The Relationship of Extreme Somatotypes to Performance in Motor and Strength Tests. *Res. Quart.* 24: 223–28.
39. SILLS, F. D. 1960. *Anthropometry in Relation to Physical Performance.* Sci. Med. Exer. Sports. New York: Harper & Row. Chap. 3.
40. TANNER, J. M. 1952. The Effect of Weight Training on Physique. *Amer. J. Phys. Anthro.* 10: 426–62.
41. ——— and J. S. WEINER. 1949. The Reliability to the Photogrammatic Method of Anthropometry with a Description of a Miniature Camera Technique. *Amer. J. Phys. Anthro.* 7: 145–86.
42. TAPPAN, N. O. and N. C. TAYGRENS. 1950. An Anthropometric and Constitutional Study of Championship Weight Lifters. *Amer. J. Phys. Anthro.* 8: 49–64.
43. TAYLOR, WILFRED et al. 1964. Conversion Scales for Estimating Humeral and Femoral Lengths and the Lengths of Functional Segments in the Limbs of American Caucasoid Males. *Human Biology* 36: 246–62.
44. THORSEN, MARGARET A. 1964. Body Structure and Design: Factors in the Motor Performance of College Women. *Res Quart.* 35: 418–32.
45. TROTTER, M. and G. C. GLESER. 1952. Estimation of Stature from Long Bones of American Whites and Negroes. *Amer. J Phys. Anthro.* 10: 463–514.
46. ———. 1958. A Reevaluation of Estimation of Stature Based on Measurements of Stature Taken During Life and of Long Bones After Death. *Amer. J. Phys. Anthro.* 16: 79–123.

F: IMPACT AND FRACTURE

1. American Institute of Mining Proceedings, (Maple Valley, Washington, August 21–24, 1962). 1963. *Fracture of Solids,* Vol. 20, D. C. Drucker and J. J. Gilman, eds. New York: Gordon and Breach.
2. American Society for Testing Materials. 1955. *Book of ASTM Standards.* Part IV: 823. Philadelphia, Pa.
3. BENNET, EDWARD, JAMES DEGAN, and JOSEPH SPIEGEL. 1963. *Human Factors in Technology.* New York: McGraw-Hill.
4. COERMANN, ROLF R. 1961. Comparison of the Dynamic Characteristics of Dummies, Animals and Man. *Impact Acceleration Stress.* National Research Council Publication 977. November: 173.
5. DAVIS, HARMER E., GEORGE E. TROXELL, and CLEMENT T. WISKOCIL. 1941. *The Testing and Inspection of Engineering Materials.* New York: McGraw-Hill.
6. DEHAVEN, H. 1942. Mechanical Analysis of Survival in Falls from Heights of 50 to 150 Feet. *War Medicine,* 2: 586–96.
7. ———. 1944. Mechanics of Injury under Force Conditions. *Mech. Eng.:* 264–68.
8. DEJUHASZ, K. J. 1949. Graphical Analysis of Impact of Bars above the Elastic Range. *J. Franklin Inst.* 248: 15, 113.

9. DEUSINGER, R. H. 1968. A Kinetic Analysis of the Back Breakfall and Force of Impact in Judo. Master's Thesis, University of Massachusetts.
10. DIETZ, A. G. H. and FREDERICK R. EIRICH. 1960. High Speed Testing Symposium 1. New York: Interscience Publishers.
11. DYE, E. R. 1956. Kinematics of the Human Body under Crash Conditions. *Clin. Orthopedics* 8: 305–9.
12. ———. 1959. Engineering Research on Protective Headgear. Symposium on Sports Injuries. *Amer. J. Surg:* 368.
13. EVANS, F. G., H. R. LISSNER, and M. LEBOW. 1958. The Relation of Energy, Velocity and Acceleration to Skull Deformation and Fracture. *Surg. Gynec. Obstet.* 107: 593–601.
14. EVANS, F. G. and H. R. LISSNER. 1960. Experimental Studies on the Relation between Acceleration and Intracranial Pressure Changes in Man. *Surg. Gynec. Obstet.* 111: 329–38.
15. ——— and L. M. PATRICK. 1962. Acceleration-Induced Strains in the Intact Vertebral Column. *J. Appl. Physiol.* 17(3): 405–9.
16. GOLDMAN, D. E. and H. E. VON GIERKE. 1961. "Effects of Shock And Vibration on Man." *Shock and Vibration Handbook.* V.3. 44–1 to 44–51. New York: McGraw-Hill.
17. GOLDSMITH, WERNER. 1961. *Impact.* New York: St. Martin's.
18. GROSS, A. G. 1958. Impact Thresholds of Brain Concussion. *J. Aviation Medicine* 29: 725–32.
19. GURDJIAN, E. S., J. E. WEBSTER and H. R. LISSNER. 1949. Studies on Skull Fracture with Particular Reference to Engineering Factors. *Amer. J. Surg.* 78(5): 736–42.
20. ———. 1950. Biomechanics: Fractures, Skull. *Medical Physics.* D. Glasser, ed. Chicago: Year Book Pub., Inc. V-2: 98–104.
21. GURDJIAN, E. S. and H. R. LISSNER. 1961. *Mechanism of Concussion. Biomechanical Studies of the Musculo-Skeletal System.* F. Gaynor Evans, ed. Springfield, Ill.: Charles C Thomas. 192–208.
22. ———, F. G. EVANS, L. M. PATRICK, and W. G. HARDY. 1961. Intracranial Pressure and Acceleration Accompanying Head Impacts in Human Cadavers. *Surg. Gynec. Obstet.* 113: 185–90.
23. HOPKINSON, B. 1921. The Pressure of a Blow. *Scientific Papers:* 423. Cambridge: Cambridge Univ. Press.
24. HUELKE, D. F. 1963. Biomechanical Studies on the Bones of the Face. Impact Acceleration Stress. Nat. Acad. Sci. — Nat. Res. Coun. Public. 977: 131–33.
25. ——— and D. H. ENLOW. 1963. Fractures of Long Bones Produced by High Velocity Impacts. *Anat. Rec.* 145(2): 243.
26. HUELKE, D. F., L. J. BUEGE, and J. H. HARGER. 1967. Bone Fractures Produced by High Velocity Impacts. *Amer. J. Anat.* 120: 123–31.
27. HUELKE, D. F. and L. M. PATRICK. 1964. Mechanics in the Production of Mandibular Fractures: Strain-Gauge Measurements of Impacts to the Chin. *J. Dent. Res.* 43(3): 437–46.
28. KENT, R. T., ed. *Mechanical Engineer's Handbook.* 1950. 12th edition. New York: John Wiley.
29. LISSNER, H. R. and V. L. ROBERTS. 1966. Evaluation of Skeletal Impacts of Human Cadavers. *Studies on the Anatomy and Function of Bone and Joints.* F. G. Evans, ed. Springer-Verlag, Heidelberg: 113–20.
30. LOMBARD, C. F. 1949. How Much Force Can the Body Withstand? *Aviation Wk.*, Jan. 17, 1949, p. 20.

31. LOVE, A. E. H. 1944. *A Treatise on the Mathematical Theory of Elasticity*, 4th ed. New York: Dover Public.
32. MANN, H. C. 1935. The relation between the tension static and dynamic tests. *Proc. ASTM* 35: 323.
33. MATHER, B. S. 1968. Observations on the Effects of the Static and Impact Loading on the Human Femur. *J. Biomechanics:* 1(4): 331–36.
34. MERRIMAN, MANSFIELD. 1915. *Mechanics of Materials*. New York: John Wiley.
35. MINTROP, H. 1957. Experimental Evolution of the Forces During Impact of a Sphere on a Plane Surface. *Z. Angew. Phys.* 9(9): 442–46.
36. NACHEMSON, A. 1961. Strength of Bones. *Nord. Med.* 65: 360–64.
37. NICKERSON, J L., A. PARADIJEFF, and H. S. FEINHANDLER. 1963. A Study of the Effects of Externally Applied Sinusoidal Forces on the Eye. Aer. Med. Res. Lab. Tech. Doc. Rep.: 63–120.
38. OKADA, K. 1956. Dynamic Studies on Bone Fractures. *J. Jap. Orthop. Surg. Soc.* 30: 105–34.
39. PERKINS, G. 1956. The Value of Knowing the Direction and Nature of the Force Causing a Fracture. *J. Bone Joint Surg.* 38B: 227–36.
40. RINEHART, J. S. and J. PEARSON. 1965. Behavior of Metals Under Impulsive Loads. New York: Dover Public.
41. ROARK, R. J. 1954. *Formulas for Stress and Strain*, 4th ed. New York: McGraw-Hill.
42. SEELY, F. B. and J. O. SMITH. 1956. *Resistance of Materials*, 4th ed. New York: John Wiley.
43. SNIVELY, G. G. 1961. Impact Survival Levels of Head Acceleration in Man. *Aer. Med.* 32(4): 316–20.
44. STAPP, J. P. 1961. Jolt Effects of Impact on Man. San Antonio, Texas: Brooks Air Force Base.
45. STUCKE, K. 1950. The Elasticity of the Achilles Tendon in Loading Experiments. *Langen. Arch. Klin. Chir.* 265(5): 579–90.
46. SWEARINGEN, JOHN J. Tolerance of the Human Face to Crash Impact. Fed. Avia. Ag., Off. Avi. Med., Civil Aer. Res. Inst. Okla., 1965.
47. TAIT, P. G. 1900. Scientific Papers, Cambridge 2: 222–49.
48. TSUDA, K. 1957. Studies on the Bending Test and the Impulsive Bending Test on Human Compact Bone. *J. Kyoto Prefect.* Med. Univ. 61: 1001–26.
49. VON GIERKE, H. E. 1961. Biomechanics of Impact Injury, Aer. Med. Lab., Dayton: Wright-Patterson Air Force Base.
50. ——— and R. R. COERMANN. 1961. The Biodynamics of Human Response to Vibration and Impact. *Rev. Med. Aer.* 2: 201.
51. WAHL, N. E. et al. 1948. *Head Impact Investigation*. Ithaca, N. Y.: Cornell Aerc. Lab., Inc.
52. WILSON, T. R. C. 1922. Impact Tests of Wood. Proc. Am. Soc. Test. Mat. 22: 55.
53. WITHEY, MORTON O. and JAMES ASTON. 1939. *Johnson's Materials of Construction*, 8th ed. New York: John Wiley.
54. ZIEGENRUECKER, G. H. 1959. Short Time Human Tolerance to Sinusoidal Vibrations. Ohio: Wright Patterson Air Force Base.

G: INJURIES AND REHABILITATION

1. Abrahams, Sir A. 1961. *Disabilities and Injuries of Sport*. London: Elek.
2. BADGLEY, CARL E. and JOHN T. HAYES. 1959. Athletic Injuries to the Elbow, Forearm, Wrist and Hand. Symp. Sports Inj. *Am. J. Surg.*, Sept: 432.

3. BEARZY, H. J. 1947. Physical Medicine in Prevention and Treatment of Athletic Injuries. *J. A. M. A.* 135: 613–16.
4. BENNETT, GEORGE E. 1959. Elbow and Shoulder Lesions of Baseball Players. Symp. Sports Inj. *Am. J. Surg.*, Sept.: 484.
5. BILLIG, H. and E. LOEWENDAHL. 1949. *Mobilization of the Human Body.* Stanford: Stanford Univ. Press.
6. BOCCACCI, C. 1957. Lesions of the Elbow Among Track Athletes and Their Cause. *Medicina Sportiva* 11: 269–74.
7. COPLANS, C. W. 1951. Lumbar Disc Herniation, the Effect of Torque on its Causation and Conservative Treatment. *S. Afr. Med. J.* 25: 881–84.
8. DE ARAUJO, H. 1949. Ocular Trauma in Sports. *Brazil Med. — Cir.* 11: 505.
9. DELORME, T. L. 1945. Restoration of Muscle Power by Heavy Resistance Exercise. *J. Bone Joint Surg.* 27: 645.
10. ———. 1951. *Progressive Resistance Exercises.* New York: Appleton-Century-Crofts.
11. DEVAS, M. B. 1958. Stress Fractures of the Tibia in Athletes or "Shin Soreness." *J. Bone Jt. Surg.* 40B: 227.
12. DEVOE, A. G. 1959. Injuries to the Eye. *Amer. J. Surg.* 98: 384.
13. ERSKINE, L. ALLEN. 1959. The Mechanisms Involved in Skiing Injuries. *Amer. J. Surg.* 97: 667–70.
14. EWERHARDT, F. and G. F. RIDDLE. 1947. *Therapeutic Exercise.* Philadelphia: Lea and Febiger.
15. FERGUSON, ALBERT BARNETT and JAY BENDER. 1964. *The ABC's of Athletic Injuries and Conditioning.* Baltimore: Williams & Wilkins.
16. FORCHER-MAYR, O. 1951. Fatigue Fracture of the Ischium in Sport Fencing. *Wien. Klin. Wschr.* 15: 331.
17. GARDEN, R. S. 1961. Tennis Elbow. *J. Bone Jt. Surg.* 43B: 100.
18. GONZÁLES, T. A. 1951. Fatal Injuries in Competitive Sports. *J. A. M. A.* 146: 1506–11.
19. HIRSCH, C. 1951. Studies on the Mechanism of Low Back Pain. *Acta Orthop. Scand.* 20: 261–74.
20. ———. 1962. Orthopedic Problems Viewed in the Light of Biomechanics. *Acta Orthop. Scand.* 32: 228–36.
21. HUDDLESTON, O. LEONARD. 1961. *Therapeutic Exercises — Kinesiotherapy.* Philadelphia: F. A. Davis.
22. JOHANSEN, OTTO. *Idrett og Skader (Sports and Injuries,* in Norwegian), State Office for Sports and Youth Work, Kronprinsensgt, 6, Oslo, Norway.
23. KENDALL, HENRY O., FLORENCE P. KENDALL, and DOROTHY A. BOYNTON. 1952. *Posture and Pain.* Baltimore: Williams & Wilkins.
24. *Kessler, Henry H.*, ed. 1950. *The Principles and Practices of Rehabilitation.* Philadelphia: Lea and Febiger.
25. KLEIN, KARL K. 1960. A Preliminary Study of the Dynamics of Force as Applied to Knee Injury in Athletics and as Related to the Supporting Strength of the Involved Musculature. *J. Assoc. Phys. Ment. Rehabil.* 14: 35–37.
26. ———. 1961. The Deep Squat Exercise as Utilized in Weight Training for Athletes and Its Effect on the Ligaments of the Knee. *J. Assoc. Phys. Ment. Rehabil.* 15: 6.
27. KRAUS, H. 1949. *Principles and Practices of Therapeutic Exercises.* Springfield, Ill.: Charles C Thomas.
28. ———. 1959. Evaluation and Treatment of Muscle Function in Athletic Injury. Symp. Sports Inj. *Amer. J. Surg.*, Sept.: 353.
29. LICHT, SIDNEY and E. W. JOHNSON, ed. 1961. *Therapeutic Exercise,* 2nd Ed. Baltimore: Waverly Press.

30. Martínez, J. L. and D. J. García. 1968. A Model for Whiplash. *J. Biomechanics* 1: 23–32.
31. McCown, I. A. 1959. Boxing Injuries. *Amer. J. Surg.* 98: 509.
32. Mercer, W. 1950. Tennis Elbow. *Practitioner* 164: 293.
33. Miller, J. E. 1960. Javelin Thrower's Elbow. *J. Bone Jt. Surg.* 42B: 788.
34. Mineo, R. and M. Zappala. 1955. The Skulls of Soccer Players. *Med. Sportiva* 9: 326–35.
35. Morehouse, Laurence E. and Philip J Rasch. 1963. *Sports Medicine for Trainers*, 2nd. ed. Philadelphia: W. B. Saunders.
36. Moseley, H. F. 1959. Athletic Injuries to the Shoulder Region. Symp. Sports Inj. *Amer. J. Surg.*, Sept.: 401.
37. Munchinger, R. 1961. Athletic Physiology: the Lifting of Weights and Dumbbells. *Jeunesse Forte Peuple Libre* 2: 1–5.
38. Norton, Martin L. and Peter Cutler. 1965. Injuries Related to the Study and Practice of Judo. *J. Sports Med. Phys. Fit.* 5: 149–51.
39. O'Donoghue, Don H. 1959. Injuries to the Knee. Symp. Sport Inj. *Amer. J. Surg.*, Sept.: 463.
40. Outwater, J. O. and M. S. Woodard. 1966. *Ski Safety and Tibial Forces*. New York: Amer. Soc. Mech. Engin.
41. Plagenhoef, S. C. 1967. Joint Moments of Force in Selected Sports and Rehabilitation Exercises. *J. Assoc. Phy. Ment. Rehabil.*, May–June: 90–96.
42. Quigley, T. B. 1959. Fractures and Ligament Injuries of the Ankle. *Amer. J. Surg.*, Sept.: 477.
43. Rasch, Philip J. and Merlin L. Brubaker. 1957. The Problem of Tennis Elbow. *J. Amer. Osteopath. Assoc.* 57: 268–71.
44. ———. 1959. Tennis Elbow: A Second Look. *J. Amer. Osteopath. Assoc.* 59: 255–67.
45. Roaf, R. 1960 A Study of the Mechanics of Spinal Injury. *J. Bone Jt. Surg.* 42B: 810.
46. Ryan, Allan J. 1962. *Medical Care of the Athlete*. New York: McGraw-Hill.
47. Slocum, Donald B. 1959. The Mechanics of Some Common Injuries to the Shoulder in Sports. Symp. Sports Inj. *Amer. J. Surg.*, Sept.: 394.
48. Soeur, R. 1963. Malleolar Fractures by Shearing. *Acta Orthop. Belg.* 29: 92–99.
49. Thomas, B. 1961. Sports Injuries Survey. Communication to British Assoc. of Sports and Medicine.
50. Thorndike, A. 1956. *Athletic Injuries, Prevention, Diagnosis, and Treatment*, 4th ed. Philadelphia: Lea and Febiger.
51. ———. 1956. Prevention of Injury in Athletics. *J. A. M. A.* 162: 1128.
52. ———. 1959. Frequency and Nature of Sports Injuries. *Amer. J. Surg.*, Sept.: 317.
53. Tucker, W. E., J. R. Armstrong, eds. 1964. *Injury in Sport*. Springfield, Ill: Charles C Thomas.
54. Waris, W. 1946. Elbow Injuries of Javelin Throwers. *Acta Chiro. Scand.* 93: 563.
55. Wessel, Janet A. and Wayne Van Huss. 1960. *Therapeutic Aspects of Exercise in Medicine*. Science and Medicine of Exercise and Sports. New York: Harper & Row.
56. Williams, J. G. P. 1962. *Sports Medicine*. London: Edward Arnold.
57. Williams, M. and C. Worthingham. 1957. *Therapeutic Exercise for Body Alignment and Function*. Philadelphia: Saunders.

Bibliography

H: KINESIOLOGY

1. ANDERSON, T. M. 1951. *Human Kinetics and Analysing Body Movements.* London: William Heinemann.
2. BADE, EDWIN. 1962. *The Mechanics of Sport.* Kingswood, Surrey: Glade House.
3. BROER, MARION R. 1960. *Efficiency of Human Movement.* Philadelphia: Saunders.
4. ——— and SARA J. HOUTZ. 1967. *Patterns of Muscular Activity in Selected Sport Skills — An Electromyographic Study.* Springfield, Ill.: Charles C Thomas.
5. BRUNNSTROM, SIGNE. 1962. Clinical Kinesiology. Philadelphia: F. A. Davis.
6. BUNN, JOHN W. 1955. *Scientific Principles of Coaching.* Englewood Cliffs: Prentice-Hall.
7. COOPER, JOHN W. and RUTH B. GLASGOW. 1963. *Kinesiology.* St. Louis: C. V. Mosby.
8. DUVALL, ELLEN N. 1959. *Kinesiology, the Anatomy of Motion.* Englewood Cliffs: Prentice-Hall.
9. DYSON, GEOFFREY H. G. 1964. *The Mechanics of Athletics.* London: Univ. of London Press.
10. FINLEY, F. RAY. 1961. *Kinesiological Analysis of Human Locomotion.* Eugene: Univ. of Oregon Press.
11. HAWLEY, GERTRUDE. 1940. *An Anatomical Analysis of Sport.* Cranbury, N. J.: A. S. Barnes.
12. HUBBARD, A. W. 1949. Methods of Research in Experimental Kinesiology. Res. Methods Appl. Hlth., P. E. Rec., Washington.
13. JENSEN, CLAYNE R. and GORDON W. SCHULTZ. 1970. *Applied Kinesiology.* New York: McGraw-Hill.
14. LEE, MABEL and MIRIAM M. WAGNER. 1949. Fundamentals of Body Mechanics and Conditioning. Philadelphia: Saunders.
15. LIPOVETZ, F. J. 1952. *Basic Kinesiology.* Minneapolis: Burgess Publ.
16. LOGAN, GENE A. and WAYNE C. MCKINNEY. 1970. *Kinesiology.* Dubuque, Iowa: Wm. C. Brown.
17. MACCONAILL, M. A. and J. V. BASMAJIAN. 1969. *Muscles and Movements — A Basis for Human Kinesiology.* Baltimore: Williams & Wilkins.
18. METHENY, ELEANOR. 1952. *Body Dynamics.* New York: McGraw-Hill.
19. MOREHOUSE, L. E. and J. M. COOPER. 1950. *Kinesiology.* St. Louis: C. V. Mosby.
20. RASCH, PHILIP J. and ROGER K. BURKE. 1963. *Kinesiology and Applied Anatomy.* 2nd ed. Philadelphia: Lea and Febiger.
21. SCOTT, M. G., ed. 1959. Research Methods. A.A.H.P.E.R., Washington, D. C.
22. ———. 1963. *Analysis of Human Motion.* New York: Appleton-Century-Crofts.
23. SLATER-HAMMEL, A. T. 1954. Two Approaches to Kinesiological Analysis. *Phys. Ed.* 9: 17–19.
24. STEINDLER, ARTHUR. 1935. *Mechanics of Normal and Pathological Locomotion in Man.* Springfield, Ill.: Charles C Thomas.
25. ———. 1962. *Kinesiology of the Human Body Under Normal and Pathological Conditions.* Springfield, Ill.: Charles C Thomas.
26. THOMPSON, CLEM W. 1965. *Kranz Manual of Kinesiology.* 5th ed. St. Louis: C. V. Mosby.
27. TRICKER, R. A. R. and B. J. K. TRICKER. 1967. *The Science of Movement.* New York: Amer. Elsevier Publ.
28. WELLS, KATHERINE F. 1966. *Kinesiology.* 4th ed. Philadelphia: Saunders.
29. WILLIAMS, M. and H. LISSNER. 1962. *Biomechanics of Human Motion.* Philadelphia: Saunders.

1: SPECIFIC SPORT ANALYSIS

Bibliography

1. ALEXANDER, JOHN F., JAMES B. HADDOW, and GERALD A. SCHULTZ 1963. Comparison of the Ice Hockey Wrist and Slap Shots for Speed and Accuracy. *Res. Quart.* 34: 259–66.
2. ALLEY, LOUIS E. 1952. An Analysis of Water Resistance and Propulsion in Swimming the Crawl Stroke. *Res. Quart.* 23: 253.
3. ARNOLD, GAIL B. 1968. An Electrogoniometric Study of the Degree of Movement of the Wrist Joint During the Golf Swing. Unpublished Master's Thesis, University of Massachusetts.
4. BARZIN, J., P. MOREAU, C. RENSON, and J. M. TILMAN. 1961. Analysis of the Flip-Flop. *Revue de L'education Physique.* 1: 49–50.
5. ———. 1961. Analysis of a Front Headspring. *L'education Physique.* 1: 157–65.
6. BELLISIMO, L. 1965. Secrets of Perfect Bowling. Englewood Cliffs: Prentice-Hall.
7. CARTER, VALERIE A. 1969. A Cinematographic Analysis of the Coast Phase in Long Jumping. *New Zealand J. H. F. E. R.*, 2: 42–50.
8. CHAPMAN, SEVILLE. 1968. Catching a Baseball. *Amer. J. Physics.* October: 868–70.
9. CLEAVER, HYTON. 1957. In Rowing. *Phys. Rec.* 9: 6–9.
10. COCHRAN, ALASTAIR and JOHN STOBBS. 1968. *The Search for the Perfect Swing.* Philadelphia: Lippincott.
11. COOPER, JOHN M. 1967. Kinesiology of High Jumping. *Biomechanics I,* First International Seminar, Zurich. 291–302.
12. COUNSILMAN, JAMES E. 1955. Forces in Swimming Two Types of Crawl Stroke. *Res. Quart.* 26: 127–39.
13. CURETON, T. K. 1935. Mechanics of the High Jump. *Scholastic Coach* 4: 9–12.
14. ———. 1935. Mechanics of the Shot Put. *Scholastic Coach* 4: 7–10.
15. ———. 1930. Mechanics and Kinesiology of Swimming the Crawl Flutter Kick *Res. Quart.* 1: 87–121.
16. ———. 1935. The Mechanics of Track Running. *Scholastic Coach* 4: 7–10.
17. DELLA, DAN G. 1950. Individual Differences in Foot Leverage in Relation to Jumping Performance. *Res. Quart.* 21: 11–19.
18. DESHON, DEANS E. and RICHARD C. NELSON. 1964. A Cinematographical Analysis of Sprint Running. *Res. Quart.* 35: 451–55.
19. DEVRIES, HERBERT A. 1959. A Cinematographical Analysis of the Dolphin Swimming Stroke. *Res. Quart.* 30: 413–22.
20. DUSENBURY, JAMES. 1968. A Kinetic Comparison of Forward and Reverse Giant Swings on Still Rings as Performed by Gymnasts with Varying Body Types. Unpublished Master's Thesis, University of Massachusetts.
21. DYSON, G. H. G. 1959. High Jumping. Amateur Athletic Association, 4th ed.
22. ELEEL, E. R. D. WILSON and C. FRENCH. 1952. Measuring Speed and Force of Charge of Football Players. *Res. Quart.* 23: 295.
23. ELFTMAN, H. 1940. The Work Done by Muscles in Running. *Amer. J. Phys.* 129: 672–84.
24. FAULKNER, JOHN A. 1966. Physiology of Swimming. *Res. Quart.* 37: 41–54.
25. ———. 1967. *What Research Tells the Coach about Swimming.* Washington, D. C.: AAHPER.
26. FENN, W. O. 1929. Mechanical Energy Expenditure in Sprint Running as Measured by Moving Pictures. *Amer. J. Physiol.* 90: 343.
27. ———. 1930. Frictional and Kinetic Factors in the Work of Sprint Running, and Work Against Gravity and Work due to Velocity Changes in Running. *Amer. J. Physiol.* 92: 583–611, 93: 433–62.

28. ———. 1931. A Cinematographic Study of Sprinters. *Scient. Monthly.* 32: 346–54.
29. FREDERICK, A. B. 1969. The Analysis of Gymnastics — A Survey of the Literature. *Modern Gymnast*, March.
30. GANSLEN, RICHARD. 1953. The Principles of Mechanics Used in Gymnastics. Unpublished Thesis, University of Arkansas.
31. ———. 1959–60. *The Mechanics of the Pole Vault.* 4th ed. St. Louis: John S. Swift and Co.
32. ——— and KENNETH G. HALL. 1960. *The Aerodynamics of Javelin Flight.* Fayetteville: University of Arkansas Press.
33. GARRISON, LEVON E. 1963. Electromyographic-Cinematographic Study of Muscular Activity During the Golf Swing. Unpublished Doctoral Thesis, Florida State University.
34. GEARON, J. 1970. A Comparison of the Golf Drive of Sanders and Weiskopf. Unpublished Master's Thesis, University of Massachusetts.
35. GROVES, W. H. 1950. Mechanical Analysis of Diving. *Res. Quart.* 21: 132–44.
36. GUSTAFSON, WILLIAM FRANK. 1955. A Mechanical Analysis of Selected Gymnastics on the Horizontal Bar, the Parallel Bar, the Side Horse, the Still Rings, and the Swinging Rings. Unpublished Doctoral Thesis, State University of Iowa.
37. HENRY, FRANKLIN M. and IRVING R. FRAFTON. 1951. The Velocity Curve of Sprint Running. *Res. Quart.* 22: 409–22.
38. HENRY, F. M. 1952. Force-Time Characteristics of the Sprint Start. *Res. Quart.* 23: 301.
39. HERMANN, GEORGE W. 1962. An Electromyographic Study of Selected Muscles Involved in the Shot Put. *Res. Quart.* 33: 85–93.
40. HOPPER, B. J. 1962, 1963. Rotation — A Vital Factor in Athletic Technique. Track Technique 9–12. *Track and Field News.*
41. HORNOF, A. 1952. The Mechanism of the Deep Jump. *Sport and Health.* Feb.: 89–95.
42. HOUSDEN, E. F. 1959. Mechanics Applied to Discus Throwing. *Discobolus, the Discus Circle Magazine.*
43. HUBBARD, A. W. 1939. An Experimental Analysis of Running and of Certain Fundamental Differences between Trained and Untrained Runners. *Res. Quart.* 10: 28–38.
44. ISHIKO, T. Personal Communication to Allen Rosenberg.
45. KARPOVICH, P. V. 1930. Swimming Speed Analyzed. *Sci. Amer.* 142: 224.
46. ———. 1933. Water Resistance in Swimming. *Res. Quart.* 4: 21.
47. ———. 1935. Analysis of the Propelling Force in the Crawl Stroke. *Res. Quart.* 6: 49.
48. KATO, H. 1960. A Cinematographic Study of Shot Put. *Jap. J. Phys. Educ.* 10: 277–79.
49. KIRKPATRICK, P. 1963. Batting the Ball. *Amer. J. Phys.* 31: 606–13.
50. KITZMAN, ERIC W. 1964. Baseball: Electromyographic Study of Batting Swing. *Res. Quart.* 35: 166–78.
51. KLOPSTEG, PAUL E. 1943. Physics of Bows and Arrows. *Amer. J. Phys.* 11: 175.
52. LESH, LONA M. 1968. A Kinematic and Kinetic Comparison of Varying Golf Drive Distances of a Woman State Champion. Unpublished Master's Thesis, University of Massachusetts.
53. *Life* Magazine. July 22, 1940. Fast Action Camera Shows How Hubbell Throws a Curve. 9: 6–8.
54. MAPES, DONALD F. 1964. Electromyographic Study of Functions of Selected Muscles in Check Action, Ballistic Movement, and Follow Through During the Baseball Throw. Unpublished Doctoral Thesis, University of Iowa.
55. MASSEY, BENJAMIN H., HAROLD W. FREEMAN, FRANK R. MANSON, and JANET A.

Wessel. 1959. *The Kinesiology of Weight Lifting.* Dubuque, Iowa: William C. Brown.

56. Mawdsley, H. 1969. A Kinematic and Kinetic Analysis of the Technique of Heading in Soccer. Unpublished Master's Thesis, University of Massachusetts.
57. Mengucci, L. 1961. Kip on the High Bar. *Educ. Fisica* 6: 1–15.
58. Miller, Robert G. and Clayton T. Shay. 1964. Relationship of Reaction Time to the Speed of a Softball. *Res. Quart.* 35: 433–37.
59. Mortimer, E. M. 1951. Basketball Shooting. *Res. Quart.* 22: 2.
60. Mosterd, W. L. and J. Jongblued. 1964. Analysis of the Stroke of Highly Trained Swimmers. *Arbeitsphysiologie* 20: 288–93.
61. Nelson, Richard C. 1964. Follow-Up Investigation of the Velocity of Volleyball Spike. *Res. Quart.* 35: 83.
62. Nett, Toni. 1961. *Die Technik beim Stot und Wurf.* Berlin: Bartels and Wernitz.
63. Plagenhoef, Stanley. 1968. A Kinetic Analysis of Running. *J. U. S. Track Coaches Assn. Quart. Rev.* October: 56–63.
64. ———. 1969. An Analysis of the Peach to a Handstand. *Modern Gymnast,* February.
65. ———. 1969 Golf Drive. *Florida J.O.H.P.E.R.,* February: 13.
66. ———. 1970. *Fundamentals of Tennis.* Englewood Cliffs: Prentice-Hall.
67. Race, Donald E. 1961. A Cinematographic and Mechanical Analysis of the External Movements Involved in Hitting a Baseball Effectively. *Res. Quart.* 32: 394–404.
68. Rehling, C. H. 1953. Analysis of Techniques of Golf Drive. *Res. Quart.* 26: 80–81.
69. Richer, M. P. 1895. Note sur la contraction du muscles quadriceps dans l'acte de donner un coup de pied. *C. R. Soc. Biol.* 2: 204–5.
70. Roberts, E. M. and Metcalfe. 1968. Mechanical Analysis of Kicking. Biomechanics I, First Inter. Sem. Zurich., 315–19.
71. Selin, Carl. 1959. An Analysis of the Aerodynamics of Pitched Baseballs. *Res. Quart.* 30 232–40.
72. Slater-Hammel, A. T. 1948. Action Current Study of Contraction-Movement Relationships in the Golf Stroke. *Res. Quart.* 19: 164–77.
73. ———. 1948. Acceleration Characteristics of the Golf Club. *Phys. Educ.* 5: 78.
74. ———. 1949. Action Current Study of Contraction-Movement Relationships in Tennis Stroke. *Res. Quart.* 20: 424–31.
75. ——— and R. L. Stumpner. 1951. Choice Batting Reaction-Time. *Res. Quart.* 22: 377–80.
76. Slater-Hammel, A. T. and E. N. Andres. 1952. Velocity Measurements of Fast Balls and Curve Balls. *Res. Quart.* 23: 95–97.
77. Stroup, Francis and David L. Bushnell. 1969. Rotation, Translation, and Trajectory in Diving. *Res. Quart.* 40: 812–17.
78. TIME. 1964. *Jai Alai.* 83: 47.
79. Tolgfors, Borje. 1958. Some Reflections on Professor Emanuel Hansen's Research Concerning Mechanical Conditions During the Breast Stroke. *Tidskrift I Gymnastik* 85: 8–9.
80. VanZeune, Guy. 1960. Study of Throwing. *Educ. Gymn.* 5: 243–56.
81. Verwiebe, Frank L. 1942. Does a Baseball Curve? *Amer. J. Physics* 10: 119.
82. Vogelsinger, Hubert. 1970. *Winning Soccer Skills and Techniques.* West Nyack, N.Y.: Parker Publishing.
83. Zimmerman, Helen M. 1956. Characteristic Likenesses and Differences Between Skilled and Non-Skilled Performance of Standing Broad Jump. *Res. Quart.* 27: 352–62.

Bibliography

1. AMAR, J. 1920. *The Human Motor.* New York: Dutton.
2. ANDERSON, M. H. and R. E. SOLLARS, eds. 1957. *Manual of Above Knee Prosthetics.* University of California, Los Angeles, School of Medicine.
3. AYLESWORTH, R. D., ed. 1952. *Manual of Upper Extremity Prosthetics.* University of California, Los Angeles, School of Medicine.
4. BARNETT, C. H. and D. HARDING. 1955. The Activity of Antagonist Muscles During Voluntary Movement. *Ann. Phys. Med.* 2: 290–93.
5. BASMAJIAN, J. V. 1961. Weight Bearing by Ligaments and Muscles. *Canadian J. Surgery.* 4: 166–70.
6. BENEDICT, J. V., L. B. WALKER, and E. H. HARRIS. 1968. Stress-Strain Characteristics and Tensile Strength of Unembalmed Human Tendon. *J. Biomechanics* 1: 53–63.
7. BIGLAND, B. and O. C. J. LIPPOLD. 1954. The Relation Between Force, Velocity and Integrated Electrical Activity in Human Muscles. *J. Physiol.* 123: 214–24.
8. BINGHAM, E. L. 1959. Fractures of the Humerus from Muscular Violence. *U. S. Armed Forces Med. J.* 10: 22.
9. BROWN, T., R. J. HANSEN, and A. J. YORRA. 1957. Some Mechanical Tests on the Lumbosacral Spine with Particular Reference to the Intervertebral Discs. *J. Bone Jt. Surg.* 39 A: 1135–64.
10. CALVIT, H. H., A. B. ROSENTHAL. 1966. An Analysis of the Structural Dynamics of the Human Body. *Proc. 19th. Conf. Med. Biol.:* 223.
11. CAMOSSO, M. E. 1958. Analysis of Mechanical Strength of Articular Cartilage Under Load. *Boll. Soc. Ital. Biol. Sper.* 34: 331–33.
12. CECI, C. O. 1949. Biomechanics of Bone. *Arch. Argent. Kines.* 2: 35–39.
13. CLARKE, H. H. and T. L. BAILEY. 1950. Strength Curves for Fourteen Joint Movements. *J. Phys. Ment. Rehabil.* April-May.
14. CLARKE, H. H. 1954. Relationship of Strength and Anthropometric Measures to Various Arm Strength Criteria. *Res. Quart.* 25: 134–43.
15. ———. 1957. Relationships of Strength and Anthropometric Measures to Physical Performances Involving the Trunk and Legs. *Res. Quart.* 28: 223–32.
16. DAVIS, PETER R. 1959. Posture of the Trunk During the Lifting of Weights. *Brit. Med. J.* 5114: 87–89.
17. ———. 1955. The Anthropometry of Body Action. *Ann. N. Y. Acad. Sci.* 63: 559–85.
18. ———. 1965. Mechanisms of Shoulder Movement. *Arch. Phys. Med. Rehabil.* 46.
19. ———. 1956. The Range of Motion of Cadaver Joints: The Lower Limb. *Univ. Michigan Med. Bull.* 22: 364–79.
20. ——— and R. F. COLEMAN. 1961. Tensile Strength of Bone Along and Across the Grain. *J. Appl. Physiol.* 16: 355–60.
21. DEMPSTER, WILFRID T. and JOHN C. FINERTY. 1947. Relative Activity of Wrist Moving Muscles in Static Support of the Wrist Joint: An Electromyographic Study. *Amer. J. Physiol.* 150.
22. DEMPSTER, WILFRID T. and G. SUZUKI. 1952. An Approach to the Localizing of Pivotal Axes of the Major Extremity Joints. *Amer. J. Phys. Anthro.* 10: 258.
23. DERN, R. J., J. M. LEVENE, and H. A. BLAIR. 1947. Forces Exerted at Different Velocities in Human Arm Movements. *Amer. J. Physiol.* 151: 415.
24. DRILLIS, R. 1958. Objective Recording and Biomechanics of Pathological Gait. *Ann. N. Y. Acad. Sci.* 74: 86.
25. EBERHART, H. D., V. T. INMAN, and B. BRESLER. 1954. The Principal Elements in

Human Locomotion. Klopsteg, P. E. and P. D. Wilson, eds., *Human Limbs and Their Substitutes*. New York: McGraw-Hill.

26. Elftman, H. 1939. Forces and Energy Changes in the Leg During Walking. *Amer. J. Physiol.* 125: 339–56.
27. ———. 1939. The Function of the Muscles in Locomotion. *Amer. J. Physiol.* 125: 357–66.
28. ———. 1944. Skeletal and Muscular Systems: Structure and Function in *Medical Physics*. Chicago: Year Book Publ. Pp. 1-20–30.
29. Evans, F. G. 1955. Studies in Human Biomechanics. *Ann. N. Y. Acad. Sci.* 63: 586–615.
30. ———. 1957. *Stress and Strain In Bones*. Springfield, Ill.: Charles C Thomas.
31. ———. 1960. Biomechanics: Stress-Strain Phenomena in Bones. *Med. Phys.*, O. Glasser, ed. Chicago: Year Book Publ.
32. ———, ed. 1961. *Biomechanical Studies of the Musculo-Skeletal System*. Springfield, Ill.: Charles C Thomas.
33. ———, ed. 1966. *Studies on the Anatomy and Function of Bone and Joints*. New York: Springer-Verlag.
34. ———. 1963. Bibliography of the Physical Properties of the Skeletal System. *Artificial Limbs* 11: 48–66.
35. ——— and M. Lebow. 1959. Biomechanical Studies on the Lumbar Spine and Pelvis. *J. Bone Jt. Surg.* 41A: 278–90.
36. Fenn, W. O. 1938. The Mechanics of Muscular Contraction in Man. *J. Appl. Phys.* 19: 165.
37. ———. 1957. The Mechanics of Standing on the Toes. *Amer. J. Phys. Med.* 36: 153–56.
38. Frost, H. M. 1966. An Introduction to Biomechanics. Springfield, Ill.: Charles C Thomas.
39. Galabov, G. 1965. Studies on the Biomechanics of the Lumbar Vertebral Column. *Izv. Inst Morfol.* (Sofiya) 11: 79–103.
40. Gaughran, George R. L. and Wilfrid T. Dempster. 1956. Force Analyses of Horizontal Two-Handed Pushes and Pulls in the Sagittal Plane. *Human Biol.* 28, no. 1.
41. Gellhorn, Ernst. 1947. Patterns of Muscular Activity in Man. *Arch. Phys. Med.* 28: 568–74.
42. Gersten, Jerome W. 1956. Mechanics of Body Elevation by Gastrocnemius-Soleus Contraction. *Amer. J. Phys. Med.* 35: 12–16.
43. Glanville, A. D. and G. Kreezer. 1937. The Maximum Amplitude and Velocity of Joint Movements in Normal Male Human Adults. *Human Biol.* 9: 197–211.
44. Govaerts, A. 1962. *La Biomecanique — Nouvelle Methode D'Analyse Du Mouvement*. Brussels: Presses Universitaries De Bruxelles.
45. Gray, James. 1959. *How Animals Move*. Edinburgh: Penguin Books
46. Haxton, H. A. 1944. Absolute Muscle Force in the Ankle Flexors of Man. *J. Physiol.* 103: 267.
47. Hettinger, T. H. and A. E. Muller. 1953. Muskelleistung and Muskeltraining. *Arb. Physiol.* 15: 116–26.
48. Hildebrand, Milton. 1960. How Animals Run. *Sci. Amer.* 202: 148.
49. Hill, A. V. 1922. The Maximum Work and Mechanical Efficiency of Human Muscles and Their Most Economical Speed. *J. Physiol.* 56: 19–41.
50. ———. 1927. *Muscular Movements in Man*. New York: McGraw-Hill.
51. Hirsch, C. and V. H. Frankel. 1960. Analysis of Forces Producing Fractures of the Proximal End of the Femur. *J. Bone. Surg.* 42B: 633–40.

52. HOWELL, A. BRAZIEIR. 1965. *Speed in Animals — Their Specialization for Running and Leaping*. New York: Hafner Publ.
53. HOYLE, GRAHAM. 1958. The Leap of the Grasshopper. *Sci. Amer.* 198: 30.
54. HOYTE, D. A. N. and D. H. ENLOW. 1966. Wolff's Law and the Problem of Muscle Attachment on Resorptive Surfaces of Bone. *Amer. J. Phys. Anthro.* 24: 205.
55. HUBBARD, ALFRED W. 1950. The Upper Limits of Slow Movement and the Lowest Limits of Ballistic Movements. Microcard PE87, Microcard Pub., University of Oregon.
56. ———. 1960. Homokinetics: Muscular Function in Human Movement. *Sci. Med. Exer. Sports*. W. R. Johnson, ed. New York: Harper & Row.
57. HUNSICKER, P. A. and R. J. DONNELLY. 1955. Instruments to Measure Strength. *Res. Quart.* 26: 409.
58. HUNSICKER, P. A. and G. GREY. 1957. Studies in Human Strength. *Res. Quart.* 28: 109.
59. INMAN, VERN T. and H. J. RALSTON. 1954. The Mechanics of Voluntary Muscle. *Human Limbs and Their Substitutes*. P. E. Klopsteg and P. D. Wilson, eds. New York: McGraw-Hill.
60. JONES, HAROLD E. 1947. The Relationship of Strength to Physique. *Amer. J. Phys. Anthro.* 5: 29–39.
61. KENDALL, H. O. and F. P. KENDALL. 1952. *Muscles, Testing and Function*. Baltimore: Williams & Wilkins.
62. KENEDI, R. M., ed. 1965. *Biomechanics and Related Bio-Engineering Topics*. New York: Pergamon Press.
63. KLOPSTEG, PAUL E. and PHILIP D. WILSON, eds. 1954. *Human Limbs and Their Substitutes*. New York: McGraw-Hill.
64. LIETZKE, M. 1956. Relation Between Weight Lifting Total and Body Weight. *Science* 124: 486.
65. LISSNER, H. R. 1961. Biomechanics Research. *J. Engin. Educ.* 52, no. 7.
66. LOCKHART, R. D. 1948. *Living Anatomy*. New York: Oxford University Press.
67. MACCONAILL, MICHAEL A. 1961. Mechanical Anatomy of Motion and Posture: Therapeutic Exercise, 2nd ed. S. Licht and E. W. Johnson, eds. Baltimore: Waverly Press.
68. ———. 1966. The Ergonomic Aspects of Articular Mechanics. *Studies on the Anatomy and Function of Bone and Joints*. F. Gaynor Evans, ed. New York: Springer-Verlag.
69. MANTER, J. T. 1949. Biomechanics of the Foot. *Anat. Rec.* 103: 486.
70. MAREY, E. J. 1895. *Movement*. London: William Heinemann.
71. MARKEE, J. E., J. T. LOGUE, M. WILLIAMS, W. B. STANTON, R. N. WRENN, and L. B. WALKER. 1955. Two Joint Muscles of the Thigh. *J. Bone Jt. Surg.* 37A: 125–42.
72. MOLLIER, S. 1938. *Plastische Anatomie: Die Konstruktive Form des Menschlichen Körpers*. Munich: Bergmann.
73. MORTON, D. J. 1952. *Human Locomotion and Body Form*. Baltimore: Williams & Wilkins.
74. MUYBRIDGE, EADWEARD. 1901. *The Human Figure in Motion*. London: Chapman and Hall. (Republished in 1955).
75. MYERS, S. J. 1968. Muscle Function and the Space Flight Environments of Weightlessness and Acceleration. Aero. Med. Res. Lab., Wright Patterson Air Force Base, Ohio.
76. PATRICK, L. M. 1961. Biomechanics New Horizons for Engineers. *Michigan Prof. Engin.* May.

77. PROVINS, K. A. 1955. Effect of Limb Position on the Forces Exerted About the Elbow and Shoulder Joints on the Two Sides Simultaneously. *J. Appl. Physiol.* 7: 387–89.
78. ———. 1955. Maximum Forces Exerted About the Elbow and Shoulder Joints on Each Side Separately and Simultaneously. *J. Appl. Physiol.* 7: 390.
79. ——— and N. SALTER. 1955. Maximum Torque Exerted About the Elbow Joint. *J. Appl. Physiol.* 7: 393.
80. RALSTON, H. J., V. T. INMAN, L. A. STRAIT and M. D. SHAFFRATH. 1947. Mechanics of Human Isolated Voluntary Muscle. *Amer. J. Physiol.* 151: 612.
81. RASCH, PHILIP J. 1954. Relationship of Arm Strength, Weight and Length to Speed of Arm Movement. *Res. Quart.* 25: 328–32.
82. ———. 1956. Effect of the Position of Forearm on Strength of Elbow Flexion. *Res. Quart.* 27: 333–37.
83. RAY, ROBERT D., ROBERT J. JOHNSON, and ROBERT M. JAMESON. 1951. Rotation of the Forearm. *J. Bone Jt. Surg.* 33A: 993–96.
84. REHMAN, I., P. R. PATEK, and M. GREGSON. 1948. Some of the Forces Exerted in the Normal Human Gait. *Arch. Phys. Med.* 29: 698–702.
85. RICCI, BENJAMIN. 1967. *Kineoenergetics. Physiological Basis of Human Performance.* Philadelphia: Lea and Febiger.
86. SALTER, N. and H. D. DARCUS. 1952. The Effect of the Degree of Elbow Flexion on the Maximum Torques Developed in Pronation and Supination of the Right Hand. *J. Anat.* 86: 197.
87. SAUNDERS, J. B., V. T. INMAN and H. D. EBERHART. 1953. The Major Determinants in Normal and Pathological Gait. *J. Bone Jt. Surg.* 35A: 543.
88. SKARSTROM, W. 1946. *Kinesiology of Trunk, Shoulders and Hip.* Springfield, Ill.: Charles C Thomas.
89. SMITH, J. W. 1957. The Forces Operating at the Human Ankle Joint During Standing. *J. Anat.* 91: 545–64.
90. TAYLOR, C. L. and A. C. BLASCHKE. 1951. A Method for Kinematic Analysis of Motions of the Shoulder, Arm and Hand Complex. Human Engineering, L. E. Abt, ed., *Ann. N. Y. Acad. Sci.* 51: 1251–65.
91. VIIDIK, A. 1966. Biomechanics and Functional Adaptation of Tendons and Joint Ligaments. *Studies on the Anatomy and Function of Bone and Joints*, F. Gaynor Evans, ed. New York: Springer-Verlag.
92. WALLACH, HANS. 1959. The Perception of Motion. *Sci. Amer.* July: 56.
93. WEIS, EDMUND B., JR. and HENNING VON GIERKE. 1964. Experimental Analysis of the Human Body as a Mechanical System. Aero. Med. Res. Lab., Wright Patterson Air Force Base, Ohio.
94. WEIS, E. B., JR. and F. P. PRIMIANO. 1966. The Motion of the Human Center of Mass and its Relationship to Mechanical Impedance. *Human Factors* 8: 399–406.
95. WHITNEY, R. J. 1958. The Strength of the Lifting Action in Man. *Ergonomics* 1: 101.
96. WILLIAMS, MARIAN and LEON STUTZMAN. 1959. Strength Variation Through the Range of Joint Motion. *Phys. Ther. Rev.* 39: 145–52.
97. WILKIE, D. R. 1949. The Relation Between Force and Velocity in Human Muscle. *J. Physiol.* 110: 249.
98. WOODSON, W. E. and D. W. CONOVER. 1964. Human Engineering Guide for Equipment Designers, 2nd ed. Berkeley: University of California Press.

INDEX

Absolute motion, 29, 47
   hand analysis, 189–90
Acceleration, angular, 47
   curve drawing, 30, 31

Balls, application of properties to sports, 83, 84
   properties, 81–83
   rebound on solid surface, 85
   spin, 84
Body rigicity, 65–69

Center of gravity, all body segments, 20, 21
   definition, 47
   sports equipment, 77–80
   trunk, 22–27
Center of percussion, 43–45
   sports equipment, 77–80
Coefficient of restitution, 82
Composite tracings, 10, 15
Computer, 6
   key punching, 171
   program, 169–85
Coriolis force, 47, 49, 50
Course outlines, 195–97

D'Alembert's principle, 33, 47
Deceleration, 39
Dempster, 18, 20, 21
Displacement, 47
   curve drawing, 29

Electrogoniometry, 4
Electromyography, 3, 38
Elevator problem, 33, 34
External forces measurement, 6, 58–61

Flight, 42
Football injuries, 71–76
Force:
   bone-on-bone, 36
   coriolis, 47, 49, 50
   external, 6, 47, 58–61
   impact, 59
   muscle, 36, 37
   one-segment motion (normal and tangential), 34–36, 47
   two- and three-segment motion, 48–50
Formulas:
   ball spin, 192–94
   coriolis force, 50
   force (one segment), 45, 46
   force (two segments), 51

Formulas: (cont.)
    force (three segments), 52, 53
    moments of (one segment), 45, 46
    moments of (two segments), 51
    moments of (three segments), 52, 53
    striking mass, 61
Free body diagram, 33
    application, 40
    definition, 47
Friction, 81–85

Glossary of terms, 47
Golf drive, 141–51
    striking mass, 61
Gymnastics, 118–20

Impact forces, 59
    football, 71–76
    handball, 66, 67
    soccer, 67, 68

Kayaking, 129–38
Kicking, soccer and football, 98–116

Link system, 89, 150

Moments:
    inertia (irregular shape), 44, 47
    interpreting, 55
    force, 37, 102
    formulas, 45–53
Motion:
    analysis of swing and impact, 77
    classification, 16, 17
    deceleration, 39
    forces, 34–36, 45–53
Motion pictures:
    advantages, 5
    camera settings, 7–10
    disadvantages, 6
Muscle:
    antagonistic, 40
    forces, 34–39
    prime stoppers, 40

Problems, 159–68
    camera settings, 8
    elevator, 33, 34
    inertial forces, 36
    static deep knee bend, 55–57
    three nonparallel forces, 32
    wood-wire, 31–32

Radius of gyration, definition, 43–45, 47
    determining, 20
    sports equipment, 77–80
Relative motion, 29, 47
    hand analysis, 190, 191
Restitution, application to sports, 82–85
Rigid body, 47

Shunt muscle, 38
Soccer, 67–69
Spurt muscle, 38
Static equilibrium:
    deep knee bend, 55–57
    three nonparallel forces, 32
    wood-wire problem, 31, 32
Striking mass during impact, 61–63
    golf, 61
    karate—boxing, 63–65
    soccer—football, 62, 101, 102
    tennis, 61, 62
Stroboscope, 5
Swimming, 121–28

Tennis serve, 138–41
Throwing, 89–97
    strength of position, 95–97
Trunk:
    center of gravity—men, 21–24
    center of gravity—women, 25–27
    moment of inertia, 24, 25

Velocity, 47
    curve drawing, 30
    football and soccer ball, 101
    kicking—foot before impact, 101

ST. MARY'S COLLEGE OF MARYLAND
ST. MARY'S CITY, MARYLAND